Regarded internationally as both a poet and a researcher, Miroslav Holub was born in 1923 in Pilsen, Western Bohemia. He studied science and medicine at Charles University, Prague, where he worked in the philosophy and the history of science departments as well as in a psychiatric ward. At that time he also began to write poetry. He earned an M.D. degree and specialized in pathology, which eventually led him to join the immunological section of the Biological, later Microbiological Institute of the Czechoslovak Academy of Science in 1958. That same year he earned his Ph.D. and published his first book of poems. Holub continues to work as a practicing immunologist. His many collections of poetry and essays have been translated and praised worldwide.

Shedding Life

ALSO BY MIROSLAV HOLUB

The Rampage

Supposed to Fly

Intensive Care: Selected and New Poems

Poems Before and After

Vanishing Lung Syndrome

The Dimension of the Present Moment

Interferon, or On Theater

Notes of a Clay Pigeon

Sagittal Section

Miroslav Holub

Shedding Life

Disease, Politics, and Other Human Conditions

Translated by
David Young

Translation assistance by
Dana Hábova, Todd Morath, Vera Orac, Catarina Vocadlova, and Miroslav Holub

MILKWEED EDITIONS

The characters and events in this book are fictitious. Any similarity to real persons, living or dead, is coincidental and not intended by the author.

Distributed by Publishers Group West

Published 1997 by Milkweed Editions
Printed in the United States of America
Jacket design by Wesley B. Tanner
Jacket and interior illustrations by Amanda Smith
Interior design by Wesley B. Tanner
The text of this book is set in Sabon.
97 98 99 00 01 5 4 3 2 1
First Edition

"Immanuel Kant" and "Truth" were first published in *Intensive Care: Selected and New Poems,* Field Translation Series 22 (Oberlin, Ohio: Oberlin College Press, 1996). Copyright © 1996 by Oberlin College. Reprinted with permission. "The Birth of Sisyphus" will also appear in *The Rampage* (London: Faber and Faber, 1997 [forthcoming]).

The epigraphs on p. vii are from Lewis Thomas, "On Matters of Doubt," in *Late Night Thoughts on Listening to Mahler's Ninth Symphony* (Viking, 1983), p. 163; and George Steiner, "Have the Arts Conceded Their Civilising Role to Science?" *The Times* (London), August 22, 1996.

Milkweed Editions is a not-for-profit publisher. We gratefully acknowledge support from the Elmer L. and Eleanor J. Andersen Foundation; Cray Research, A Silicon Graphics Company; Dayton's, Mervyn's, and Target Stores by the Dayton Hudson Foundation; Ecolab Foundation; General Mills Foundation; Honeywell Foundation; Jerome Foundation; The McKnight Foundation; Andrew W. Mellon Foundation; Minnesota State Arts Board through an appropriation by the Minnesota State Legislature and in part by a grant from the National Endowment for the Arts; Creation and Presentation Programs of the National Endowment for the Arts; Norwest Foundation on behalf of Norwest Bank Minnesota, Norwest Investment Management and Trust, Lowry Hill, Norwest Investment Services, Inc.; Lawrence and Elizabeth Ann O'Shaughnessy Charitable Income Trust in honor of Lawrence M. O'Shaughnessy; Piper Jaffray Companies, Inc.; Ritz Foundation; John and Beverly Rollwagen Fund of the Minneapolis Foundation; The St. Paul Companies, Inc.; Star Tribune/Cowles Media Foundation; James R. Thorpe Foundation; and generous individuals.

Library of Congress Cataloging-in-Publication Data

Holub, Miroslav, 1923–
Shedding life : disease, politics, and other human conditions / Miroslav Holub ; translated by David Young . . . [et al.]. — 1st ed.
p. cm.
Includes bibliographical references.
ISBN 1-57131-217-X
1. Science—Popular works. 2. Medicine—Popular works. 3. Ecology—Popular works. I. Title.
Q162.H767 1997
570—DC21 97-10712
CIP

This book is printed on acid-free paper.

Shedding Life

ANGELS OF DISEASE

TROUBLE ON SPACESHIP EARTH

We can take some gratification at having come a certain distance in just a few thousand years of our existence as language users, but it should be a deeper satisfaction, even an exhilaration, to recognize that we have such a distance still to go.

LEWIS THOMAS

I remain unrepentant in my hunch that intellectual energies, imaginative boldness and sheer fun are currently more abundant in the sciences than they are in the humanities. Courteous inquiries by colleagues in the sciences render even more embarrassing the casuistic jargon, the pretentious triviality which now dominate so much of literary theory and humanistic studies.

GEORGE STEINER

Shedding Life

Angels of Disease

Immanuel Kant

The philosophy of white blood cells:
this is self,
this is nonself.
The starry sky of nonself,
perfectly mirrored
deep inside.
Immanuel Kant,
perfectly mirrored
deep inside.

And he knows nothing about it,
he is only afraid of drafts.
And he knows nothing about it,
though this is the critique
of pure reason.

Deep inside.

Miroslav Holub
translated by David Young and Dana Hábova

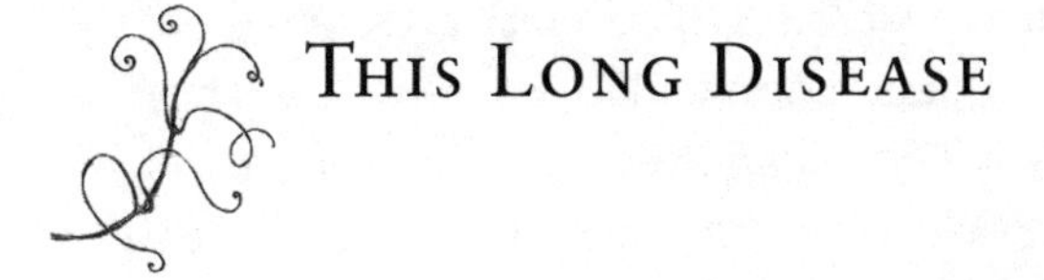

This Long Disease

We are deeply convinced that health (physical and mental) is an independent reality, one that overlaps with our concept of life and is its crowning biological aesthetic and moral value. Through the development of civilization, the belief goes, we have achieved a higher level of health for the individual, who lives longer and more contentedly than before, and enjoys a better chance of mental stability. Our improved level of health is our main argument for the advantages of Western civilization, and we take the lengthening of the human life span as the main proof that we aren't going round and round in a vicious circle.

Moreover, because health is defined only as the absence of disease, we automatically accept the scenario in which we wade out of the darkness, mire, and dirt, away from the power of worms and invisible demons, and thereby become ourselves: people who are rosy, healthy, unthreatened, and collectively productive into an old age that was once the privilege of exceptional individuals.

We accept the dichotomy that makes the concept "self" antithetical to the concept "disease," meaning "that aggressive pathological nonself." In Schelling's romantic terms, the disease is the disharmony of organs, and we must get over this disharmony, in order to return to the original unity of human being and consciousness. In fact, of course, this dichotomy doesn't exist. The angel of disease is kith and kin, identical with the historical phenomenon of people. There wouldn't be any people if there were no evolutionary pressures from disease and death, degeneration and loss of function. The multicellular organism is based on the capacity of some of its cells for self-sacrifice. We are just as much the result of diseases and constant tiny deaths as we are the result of a fundamental tendency to preserve the permanence of an organism's inner environment: we are the result of changes in genetic material over the course of a million years, just as we are the result of its relative constancy in the measures of our history. We are also, perhaps, the result of retroviruses copying themselves into the one fundamentally human thing that we have, the cell nucleus, as well as the product of slow viruses attacking the nucleus of one or another cultural belief.

I bring up slow viruses just to be on the safe side. The kuru syndrome in the closed population of New Guinea and the fact of Alzheimer's disease in the open tradition of Europe and America are just two unpleasant reminders of how, in history, sometimes at a distance (kuru) and sometimes close to home (Alzheimer's), a presence as dark as a slow virus can change the outline of a human being and interfere with familiar adages about the wisdom of elders.

A retrovirus is the polar extreme case of a disease, an embryo of a disease that is trying to take over the very essence of the host, let's say the very essence of what we see as a person's health and identity, for its own dark or light goals. Retroviruses, then—and of course this includes HIV, the cause of today's AIDS and possibly previous epidemics—were not angels of death but angels of takeover, angels of temptation and disturbers of the human genome. From the human perspective, therefore, they are the creators of increased variability in the species. Some of the retroviruses from ancient history are written into our genomes today. It's no coincidence that some of our genetic components and intercellular signals have nucleotide sequences very similar to nucleotide sequences of retroviruses. For example, something as basic as the reproduction of connective tissue cells (fibroblasts) is triggered by a gene that is noticeably similar to a retrovirus sequence. When we look at the complicated, branching diagrams of cell interactions and intercellular signals, we see not only the logical record of an organism's inner evolutionary drama, but also the record of more or less accidental penetration by the alien nonself, which was later appropriated and productively used, but which complicated the neatness of the dramatic line. The human drama may have started with the mitochondria that every properly nucleated cell uses as an energy source, a power point, even though these mitochondria may have originated as symbiotic bacteria that gave up their own evolution.

The fact that microbes and viruses—during the two billion years that multicellular organisms have been in existence—and retroviruses—in the four million years of the evolution of the human line—didn't take us over must be ascribed, for the most part, to the prudence or frivolity of the demiurge who had the preservation of the biological species in his job description. Or, more simply, to the facts that a genome can't be transferred that readily and that a species can't be exterminated easily—or at least not quickly—through biological mechanisms. In every population, and in the case of every pathogen, even if it attacks areas of the body as central and sensitive as helper T lymphocytes in the case of the AIDS agent, somebody will survive; even, as Nietzsche reminds us, become stronger. Or at least the surviving population is strengthened. In the case of HIV infection it would be the bearers of the genomic defect leading to the lack of expression of helper T-cell receptors.

In any case, the natural substance of a human being is to a certain extent formed from retroviruses, which, incidentally, in destroying their host destroy their own future potential, so that they survive only when they don't succeed in destroying the host. We are a genetic chronicle, a good fifth of which is written in absolutely primitive viral syntax.

If we want to continue to use the concept of disease, then we should recognize the direct participation of disease, and its causative agencies, in the human process. But there is even more to the case than that; other powerful, if indirect, mechanisms associated with disease are at work in shaping us too.

The body's immune system, the guarantee of its uniqueness and inviolability, is influenced and shaped by extrinsic disturbances, which thus, in the historic sense of the word, implant their character into the body's own, hereditary substance. The surfaces of our cells, for example, share antigens with some microbes, and so an infection can even lead to a cross-reaction against the body's own components: in the historical sense, to the elimination of certain markers and their carriers. This includes even such obvious features as blood groups in the A, B, and O system and the number of their carriers in a certain population.

The gene or genes for hereditary deviations and illnesses survive in some populations because they also bring some resistance to some prevailing infections. Metaphorically, the cause of one disease protects against another. A classic example is sickle-cell anemia. It is fatal in homozygotes that have both the father's and the mother's gene for the deviant hemoglobin structure. Heterozygotes, with just one gene, don't develop the disease and are protected against malaria caused by *Plasmodium falciparum*. The selective advantage for the population preserves a gene that is detrimental for some individuals.

The proper forces of the so-called healthy body may happily participate in a process that would generally be perceived as illness. This can happen not only through mistakes of the complicated regulatory mechanisms of immune reactivity that lead to attacks by our own lymphocytes against our own selves; it also happens during the completely flawless functioning of our defenses. Antibodies against some viruses (including HIV) can paradoxically enhance the progress of the infection. Severe organ damage is caused in some circumstances by antigen-antibody complexes. Despite the good intentions of the antibody molecules, grave shock states occur not through direct action of the bacterial endotoxins, but through the general alarm of the organism sensing their presence. Pain, swelling, and irreversible damage are produced by the body's own mediators of inflammation, and the heavy artillery of bodily defenses, the phagocytic cells, find their most effective ammunition in free oxygen radicals that destroy not only the criminal microbes, but the body's own tissues as well. Friendly fire, as they say. As Peter Medawar stated in his *Future of Man,* the remedy *is* the disease.

Further, the infections that were capable of causing the greatest destruction in our history, infections such as plague, smallpox, syphilis, anthrax, tuberculosis, and leprosy, evoked substantial changes in the very nature of our immune systems. We are the descendants of people who survived epidemics in the cities of the Middle Ages. We are probably the result of the natural selection of immune systems that were effective against certain infections and against foreign agents in general. Let's say we are descendants of people with "strong," highly reactive immune systems. So it's no wonder that we are also a population with a tendency to a generally "high" reactivity, including immune reactions against the body's own components, a tendency to autoimmune reactions. Type I diabetes could be a contemporary consequence of having passed through the ancient tunnels of various plagues.

That's how the fifteenth and sixteenth centuries affect today's medical problems, and that's how the angel of disease shows up in our health today. If we investigate the history of disease, we also investigate the foundation and future of health. Each of us tends to believe that disease is the punishment of others, something that doesn't affect us because we live morally and hygienically and use antibiotics: we accept leprosy and smallpox as the scourges of uncivilized filthy heretics, and syphilis and tuberculosis as the remnants of a time when poverty was rampant and medicine wore a dinner jacket instead of a sterile lab coat. In reality, however, in our shape and essence we are to a certain extent the fruit of these diseases, and future generations will to a certain extent be the fruit of diseases that are becoming epidemiologically dominant today. Of course this does not only mean infectious diseases, though new ones are appearing, even in the age of antibiotics, but also degenerative and metabolic diseases, which gain ground to the same degree that the human life span is extended.

Not even the antibiotics era presents a final solution: modern medicine is not the only player on the chessboard of infectious diseases; microbial genetics is also at play. Every move of the human intellect encountered a counter move in bacterial chromosomal changes or in the exchange of the genetic material between microbes. The microbe even received unwitting assistance from general practitioners and clinicians for whom penicillin became the philosopher's stone and a remedy for almost any sore throat. The rigidity of human reason was no match for the plasticity of microbes. The penicillins have become practically a placebo.

A highly prevalent disease, in the periods of its maximum development, influences human typology. It becomes, at least at certain social levels, a somatic and psychic standard, not just a pathology but also a kind of cultural physiology, as was the case with tuberculosis in the last century. Its carriers appeared to exemplify artistic bohemianism and intellectual intensity, as Susan Sontag has perfectly demonstrated in her book *Illness as Metaphor.* The illness became the sign of the heightened uniqueness of its carrier, and if it also had, like syphilis, psychiatric consequences, then it meant increased sensibility, transcendence of self, and sometimes creativity. According to ancient religious myth, a disease eating away at the body has a tonic effect on the soul: the more granulomas on the surface, the more visions inside.

An illness that strikes artists—venereal disease, for example—can influence their spiritual tempo, their illusions and delusions, and finally become a style-creating factor. A lot depends, of course, on whether the psychopathological consequence of the given illness unfolds in New York or in Peoria.

The epidemics that decimated entire city populations in Europe in the fifteenth and sixteenth centuries, and analogous earlier epidemics, about which we know less, radically changed not only ways of life, but also ways of faith. Syphilis, for example, made "vice" and "sin" visible in princes and prelates and led to attempts at regulating prostitution; it also led to "healing" practices that resembled the tortures of hell.

Stanislav Andreski, in his book *Syphilis, Puritanism and Witch Hunts,* goes so far as to argue that the spread of syphilis and other sexually transmitted diseases in Europe resulted on the one hand in the growth of puritanism and radical reformist movements within the sixteenth- and seventeenth-century church, and on the other hand in the mass spread of demonology. Deformities and states of dementia in the late stages of venereal disease and in congenital syphilis were probably seen as proof of witchcraft and sorcery. Paradoxically, the spread of manuals on witch-hunting benefited from the printing press, that product of reason. Ironic that inventions of reason so often in history serve as carriers of fate.

Infections changed the appearance of cities, from manorial residences to hospitals and jailhouses. Syphilis, in its late or congenital form, had a long-lasting impact on the psychology of the citizen and the psychology of creativity. As late as 1915, Léon Daudet believed that:

> treponema is a force as much behind genius and talent, heroism and wit, as behind general paralysis, tabes and almost all forms of degeneration. This microbe carries manias and hemorrhages, great discoveries and great scleroses, is strengthened through the marriages of relatives in affected families and plays a role comparable to the role of Fate in classic antiquity. It makes a great poet out of a maid's son, a satyr out of a peaceful citizen, and an astronomer or conqueror out of a sailor. An age like the sixteenth century . . . looks like the invasion of treponema into the elite and the masses, like the sarabande of congenital syphilis (heredos). From the first line of his famous dedication Rabelais was clear about the essence of the matter, and was doubtless himself a direct participant and victim, as his dazzling language and the swirling of his excited, brilliant pictures attest. Treponema nourishes the dramatic intensity of life and is at the same time its curse.

Although it doesn't sound very uplifting, the basis of the contemporary human being is derived not only from cultural traditions but also from the history of human diseases, which themselves left a pretty strong imprint on cultural traditions and, in addition, influenced our physiology, our immune capacity, and our identity.

The history of life is the history of its being endangered, the endangering of life by life, as much as it is the development and strengthening of life through danger. We can describe human history equally well as a long development or as a long disease. Humanity can echo what psychiatric patients often say:

What would I be without my disease?

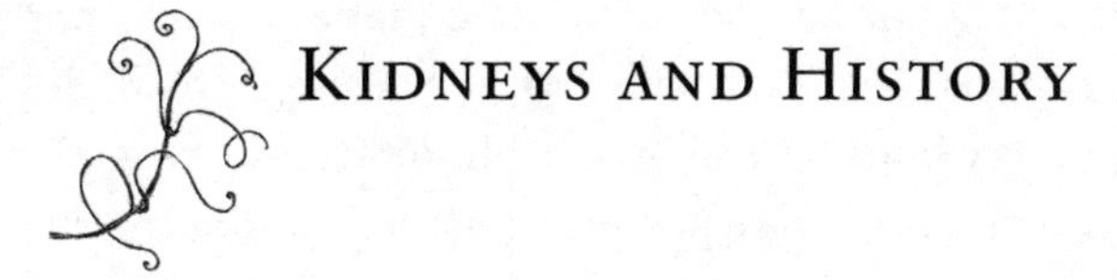

Kidneys and History

In history—or, more precisely, in the recorded life stories of distinguished citizens—it is the failing organs, rather than the fully functional ones, that figure and become famous. Among the failing organs of history, the kidneys hold an honorable place.

Thus, in 1547, what defeated the French king Francis I in his lifelong struggle with Charles V over who was to have the deciding word in Europe was not syphilis, as would have been appropriate and as Rabelais implied in his verse about the king dying in an unbearable, putrefying stench, but septicemia—colloquially, blood poisoning—on top of chronic illness of the urinary tract, abscesses, and fistulas, of either tubercular or gonorrheal origin.

Of course, there are no certain diagnoses for that century, or, for that matter, the ones that follow it. If we describe it as the early modern era, we are talking about political and military technology rather than about medical technology. For disease and illness there are only legends, more or less plausible, to give humanists the sense that they are bringing some concrete science into historical study.

This is the area in which we find the noble story of the kidneys and urinary tract of Tycho Brahe, who died in Prague in 1601 because "he didn't dare urinate." Milan Kundera classifies that as comical immortality. Nonetheless, whether death was the result of prostate adenoma, syphilitic tabes, or inflammation of the nerves in advanced diabetes, whether it was urosepsis from progressively infected kidneys or a tumor in the lower pelvis, it could hardly have been simply from not daring to pee: the bladder will hold around three liters of urine. We can ask a prostate patient or diabetic about the accompanying sensations. Immortality may cheer him up somewhat, even if it won't make him laugh.

An especially renowned set of kidneys and urinary tract belonged to the lord protector of England, Oliver Cromwell. Their renown is mainly attributable to Blaise Pascal. In Pascal's *Pensées* we read:

> Cromwell was about to lay waste to the entire Christian world: The royal family would be destroyed, and his own achieve power for all eternity, were it not for the bit of sand that got into his urinary tract. Rome itself might have quaked before him, but that little stone got in the way. He's dead, his family is humbled, peace is everywhere, and the king is restored.

Cromwell, a religious fanatic and military leader as ruthless as he was intelligent, instigated Europe's first revolution, which ended with the execution of Charles I in 1649. Thanks to his kidney trouble, Cromwell entered the history of political pathography and pathology through the philosopher's pen.

Pascal's elegant discourse contained, as sometimes happens, one error: Cromwell died of typhus during the epidemic that raged in London and all of England in 1658. Pascal's deadly little stone is the historic document of a false diagnosis.

Blaise Pascal himself, that great spirit of the French intellectual pantheon, didn't spend a single day over the age of eighteen without unbearable migrainelike pain, perhaps from an aneurysm in the lower skull, depressing evidence of the influence of a great disease on a great mind (an attribute most rulers can't complain of). Even as a boy, Pascal was already equipped with the intellectual brilliance with which one enters the history of science. On his own initiative, just for himself, he discovered and formulated anew some of Euclid's theorems. And he described with precision some phenomena of acoustics and mechanics. Plagued by pain, and with difficulties in the simple ingestion of food, he later turned to mysticism, from the spirit of geometry to the gentle spirit that—according to Pascal's definition—sees directly into things, and from that to the pure feeling of devotion to the God of love, who stands above the God of truth. His prayer to God makes good use of his afflictions: illness is "the Christian's natural state" and an instrument of salvation, for it frees us from the world of vanity and leads us to God's mercy.

According to a new French theory, Pascal's difficulties, which include the conspicuously early deaths of his mother, his sister, and the children of their families, right up to his convulsive end, can be explained by the cystic degeneration of the kidneys, a hereditary illness often connected with arterial aneurysms in the brain and other organs.

Cystic degeneration of the kidneys wasn't recognized until 1888, so the diagnosis of Pascal's suffering couldn't even be wrong, because it wasn't yet possible. The doctors simply sped up his end through massive induction of liquids that caused "great vertigo and great headaches"; he drowned in his own water during complete kidney failure.

Even the kidneys' entry into the more rational pathology of internal diseases didn't diminish their historical role. Gout, kidney stones, hydronephrosis, and chronic kidney failure caused Emperor Napoléon III to gradually fall apart, physically and morally, from 1863 on, and by 1866 it seemed to Bismarck that he was an incompetent human ruin. Had he listened to his doctor and undergone a timely operation,

Napoléon might have changed the history of Europe in 1870. On July 14 of that year he had to leave the ministers' council because he felt unwell, and during his half-hour absence the Empress Eugénie compelled the council to declare war on Bismarck, which led to the French armies' quick collapse on September 2. Napoléon III had to entrust their leadership to the mentally ill Marshal Bazaine because in the field he himself was inclined to hide so that the troops wouldn't notice that his face had to be painted with makeup and that he had to be lifted up onto his horse. He died of uremia in 1873, after three operations, in a chloroform narcosis that probably finished off his kidneys.

Some quite recent history-making kidneys belonged to the liberator and none-too-coherent president of Indonesia. In 1965, twenty years after the Indonesian declaration of independence, Sukarno ignored the advice of a Viennese nephrologist who advised the removal of a kidney, probably because of a benign tumor. Instead, he persevered in hectic political activity and gambled on a communist overthrow, an event that began and ended in a legendary massacre. Sukarno was replaced by President Suharto so that he could devote himself fully to an elegant and ideologically more acceptable cure of kidney ailments, namely, Chinese acupuncture. While manifesting completely bizarre behavior, he died in 1970 of "the results of kidney illness and ensuing complications." How Southeast Asia might have changed had Sukarno undergone an operation at the right time, we can only guess.

The Algerian president Houari Boumedienne was stricken with kidney failure and a suspicious paralysis of the facial nerve in 1978, at the age of forty-six. For reasons of secrecy he left on a six-week "official and friendly visit" to Moscow, where, in diplomatic dusk, the medical cause was supposed to be discovered but was not. After his return home, Boumedienne fell into a coma, and only then did the French doctors who were summoned discover advanced Waldenström syndrome, with complete kidney failure, which the hastily summoned Waldenström himself confirmed a few days later; the powerful diagnostic instruments immediately delivered from the USA and West Germany showed several successive strokes. The kidney functions stabilized after dialysis, but in two weeks further circulatory problems and internal bleeding appeared, and the president was kept alive until a politically convenient time only with the help of machines. A consortium of up to seventy doctors from around the world, along with technology delivered by giant planes, couldn't make up for the time lost before the correct diagnosis, and Boumedienne died, a clear example of a situation where illness doesn't influence political history but politics influences the illness of a statesman. A statesman's right to professional medical confidentiality is only relative anyway; a statesman's illness touches everyone, sometimes very specifically. Besides, it's not the doctors who decide about confidentiality, it's the interests of the state.

Failing kidneys are also assured of an honorable place in the history of the USSR, where experts in the KGB, the NKVD, and their predecessors easily diagnosed and treated the cerebral arteriosclerosis that struck Lenin at the early age of forty-seven. (The impressively large nerve cells that distinguished Lenin's brain after death turned out to be a pathologist's mishandling of the brain specimens.) The experts dealt with Stalin's paranoid state in 1952–53, with Khrushchev's somewhat psychopathic character, and with Brezhnev's myocardial attacks, high blood pressure, and heavy, practically Leninist cerebral arteriosclerosis. When the sixty-eight-year-old Yuri Andropov assumed the leader's post, it was officially if vaguely acknowledged that he wasn't exactly in the best of health. This was something of a breakthrough in politicodiagnostic practice, according to which a statesman is at the peak of his powers as long as he's not in a mausoleum. Andropov assumed his post in a state of health in which less illustrious citizens are shipped off to health-care storage facilities because there's no room for them, ordinary mortals, either on the artificial kidney machine or on the waiting list for a combined pancreas-kidney transplant. He had advanced diabetes and related kidney disease, for which he underwent dialysis three times a week. He suffered from diabetic illness of the retina and could read a report only if it was written in huge letters; furthermore, there were multiple inflammations of the nervous system. On top of that he had Parkinson's disease, and despite L-dopa treatment, a few months after assuming power he could no longer move without being held up by his bodyguards. Andropov was under the long-distance care of Helsinki doctors, but even in Moscow his indisposition became obvious, and it was finally acknowledged to be serious (something had to be said when he disappeared for a few months at a time). The press releases called the nephrosclerosis a cold and described the Parkinson's disease as a case of the flu. But at least we weren't told he was at the peak of his powers as a statesman. It must have been decided that it was dangerous to joke about the kidneys.

The official diagnosis, the day after his death, was actually extensive and truthful. It even included "progressive hypertension" and considerable "dystrophic changes in the internal organs," undoubtedly caused by long-term dialysis as much as by diabetes. Time for the transplant, which was supposed to be performed by German specialists, just ran out.

Rulers and their kidneys can, to a certain degree, be replaced and forgotten, but I recall one kidney infection, glomerulonephritis, that had cultural, and thus irreparable, results. What might a certain man have written, had he lived to the age of fifty and not merely thirty-five, a man who wrote three great symphonies in three months, four operas in five years, and thirty chamber compositions a year at a time when he already had chronic uremia and high blood pressure? A progressive swelling is obvious even in idealized portraits.

Even as a child prodigy, Wolfgang Amadeus Mozart, according to new documentation, had a continuous series of focal infections, two attacks of rheumatic fever, erythema nodosum, smallpox, and typhus. At fourteen he lived through glomerulonephritis; at twenty-eight he apparently contracted an ascendant purulent kidney infection with episodes of colic, and his kidney functions definitively began to fail. By today's criteria he should have been on a disability pension, while in reality it wasn't until 1787, at the age of thirty-one, that he attained the secure position of court composer and emperor's chamber musician, which function he fulfilled with enviable productivity and energy. Only two years before his death did his compositional activity decline. Periods spent languishing in bed alternated with short episodes of feverish composition, travel between Vienna and Prague, and even a stint as conductor of *The Magic Flute*. By today's medical standards, Mozart would have written *La Clemenza di Tito* and the *Requiem* on dialysis, while awaiting a transplant.

The historic losses caused by kidneys are numerous, even without dragging in philosophic scriptures, the Middle Ages, and similar dark times before the dawn of medical recognition, or after it. The losses caused by the acute and chronic failure of organs that are less politically unique, like the brain, for example, are less obvious. This is especially true in government organizations, where the pressure of administrative (natural) selection gives an advantage to brains that are relatively less functional, sometimes indeed merely storage boxes.

Which can't be said about kidneys.

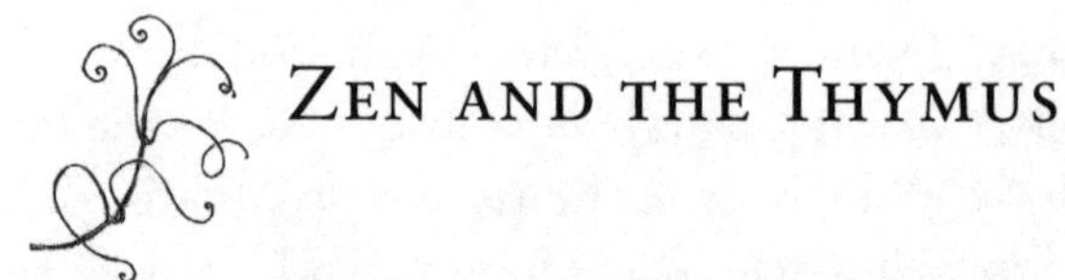

Zen and the Thymus

Most normal human beings know what Zen is, at least in terms of geographical origins and spiritual depths, but don't know anything about the thymus. In terms of the body, we delimit our geographic attention to the visible organs and our sense of depth, spiritual or otherwise, to culturally essential organs like the heart, lungs, and genitalia. We consider the thymus a sort of philosophical frill, a bit of background noise above the heart and behind the sternum, a vaguely outlined clump; it's part of a negligible anatomical background with a foreground consisting of noble thoughts on the essence of being, sometimes mixed in with some thoughts about having a good time or lining up at the trough.

Despite this poor reputation, the thymus happens to represent the real brain of immunity, and hence of survival in this aggressive natural world. It's in the thymus that the units of the body's inner wisdom, the white blood cells, and lymphocytes, learn what to react against and what to disregard or tolerate, so that the body and the spirit maintain themselves and don't succumb to foreign intrusions or to impulses of self-alienation. Immunology was built on the basic assumption that lymphocytes reacting against the self either didn't exist or must be eliminated somewhere in the body (precisely in the thymus, it turns out), while lymphocytes recognizing and reacting against antigens from the outside, nonself antigens, must somewhere in the body be approved of, praised, and corroborated (precisely in the thymus, it turns out).

Which sounds a little bit like "to study yourself is to learn to forget yourself," as the Zen masters tell us.

But let's stick with immunology for the moment. The process of negative and positive selection in the thymus is difficult to prove and is in any case a leaky process. All too often cells that react against the self survive and eventually trigger a disorder known as autoimmune disease, such as diabetes type I or lupus erythematosus. (There are four hundred others.) Second, there is a logical possibility that the immune system, especially in early fetal development, is responsible for the control and integration of all tissues and organs, all the building and restructuring that consist of recognizing and grinding down decaying cells and modulating the proliferation of others. The system is not blind to the body's own antigens. It is used to them and builds sophisticated control mechanisms to limit response to these antigens to the right intensity and duration. Again, of course, the control mechanisms are not infallible. In this sense, the immune system could be said to have its own more or less accurate internal vision of the entire organism, an internal image in which the tissues needing more immune attention are represented more emphatically than others—so that the mobile mind of immunity imagines a sort of homunculus in the brain also. Nonself antigens provoke immune responses of a higher intensity if they closely resemble self antigens. The likelihood and extent of immune reactions is demonstrable on microbial antigens; we react best against those that are developmentally so ancient that they share characteristics with our own cells. *Mycobacterium tuberculosis* is a notable example. Thus it is that the immunological homunculus may have his say even in cultural history, as in the last century, when tuberculosis became a trademark of spiritual sublimity.

In a Zen interpretation, this change in immunological thinking might be phrased thus: "to study the foreign is to remember oneself."

We can say of immunology, as of any contemporary science, that its mind is anxious because it is in a state of permanent doubt about itself. It is a kind of thinking that takes place on pins and needles. Thinking that doesn't doubt itself is a lot easier and more satisfactory; it is backed by traditions and is sometimes thousands of years old, very agreeable and pleasing because it can be grasped by a certified idiot as readily as by a certified wise man. It is a thinking wrapped in cotton wool.

The first kind of thinking operates with more or less defined and more or less comprehensible terms, and its moves toward definition require prolonged and even painstaking study; the second kind operates with soft, featherweight terms, ones that we know from the fairy tales we used to hear before falling asleep.

Let us see how thymic events would be described by these different kinds of thinking and speaking.

The precursors of thymic lymphocytes (T cells) travel during fetal development into the thymic anlage, a tiny spongelike structure composed mainly of epithelial cells, in two waves. In the thymus the lymphocytes then either undergo the preprogrammed cellular death or are saved from it: conventional T cells that have rearranged their alpha and beta genes for the antigen receptor (TcR) are controlled by a certain TcR beta protein that triggers in them the expression of suppressor markers (CD8) and helper markers (CD4), as well as the transcription of the alpha locus for TcR and the expansion of their likes, their clone. At later stages of the intrathymic passage of alpha/beta T cells, it is their specificity (fine recognition of antigens) that decides further. They proceed into their preprogrammed death if they do not find something to combine with. They are instantly destroyed if they meet the marker of self plus a special polypeptide, or they are rescued from death by binding to the self-marker expressed on a thymic epithelial cell and combined with another polypeptide, one that provokes an increase in the density of the T cell's surface receptors and markers and thus determines whether the cell will go to the suppressor or the helper pathway upon emigration from the thymus. (And all that only in the case of conventional T cells, without consideration of the intriguing gamma/delta type, so unconventional that they may develop even without a thymus.)

All of these steps include a number of even less comprehensible substeps, and each substep includes a vast number of extracellular and intracellular events that would fill, in a biochemical simplification and graphic representation, a blackboard the size of the main gate of Hagia Sophia.

And all that would be simply a metaphoric expression of the eventual reality, a mind's way of producing order from the chaos. And by next week some of the metaphors will be discarded and replaced by others. The only consolation a biologist has is that a semiotics scholar looking at the thymus would express it in many more incomprehensible terms and sentences.

So now let us render the thymic story in the tender, plastic, all-embracing language of Zen, in which no uncertainties exist: The thymus, not unknown to Gautama, is simply a part of our soul. The aim of the soul is to recognize itself. Only through its imperfect existence may we recognize perfect existence, which exists in each of us in a rudimentary form, just as the ocean is present in each of its waves. Our life and our death are one and the same thing. To grasp that requires that we rid ourselves of all preconceived notions. Only thus can we cross to the other shore. Even the smallest particles of ourselves long to make this crossing. As the great thirteenth-century Zen master Dōgen said, the crossing is nonattachment and the nonattachment is a kind of giving. The smallest particles do not attach, but move, wave after wave, to the post-Zen space where they either find their peace or try to give of themselves, moment after moment, wave after wave. At a certain stage of giving there is nothing that is unknown to the givers. Their way is "kshantee pradjna paramita" or persistence in seeking the state where the self will expand, by great experiences and encounters, into the "great self," mirroring some part of the sky. The expansion through great experiences is the aim of their activity and of their spirit. In one sense, their experiences are always new. In another sense, however, they are reincarnations of the great spirit and they possess its vision. Nevertheless, they concentrate on everyday life, because Zen is a deepening of the obvious and the everyday. The self particles will be lost, if they are taken and carried away by their own change and overwhelmed by their own anxiety. They will be saved if they concentrate on their own silence in this noisy world. Their spirit will embody the strengthening of the steadfast and genuine belief in the identity of what you give and what you are, the strengthening of your self. There will be two kinds of self particles, those on the route to yin and those on the route to yang.

I am quite open to the idea that expressions and terms like "T-cell receptor," "transcription," "gene rearrangement," "polypeptides," "epithels," and "self-markers" are more repulsive, unnatural, and alien to the human mind (even though I'm not sure what that is) than "nonattachment," "giving," "other shore," "great spirits," "waters," "sky," "irritation," "deepening," and "silence," not to mention "kshantee pradjna paramita," which is so much more human than "CD8" and "body-scanning suppressors."

I would only suggest that in both interpretations we remain ignorant of what goes on in a thymus or elsewhere in the body of either a senior immunologist or a master of the koan. It's just that the immunologist is prepared, after twenty years of personal effort and in view of a hundred thousand papers published so far, to accept and acknowledge his relative ignorance. He knows the difference between a metaphor and a word escapade.

Venerable masters are prone to jump to the conclusion that a thousand (or even a hundred) years of belief constitute an argument, that the old words are much more human than the new ones. Masters are masters because they have learned how to be sure in the absence of any real reasons for certainty. Masters are the function of a culture in which jumping to conclusions is an essential part of the ritual. Immunologists are the function of a culture in which jumping to conclusions means endangering the survival of men, women, and infants.

In view of AIDS, diabetes, myasthenia gravis, and organ transplantation, I prefer the immunologist's thymus to Zen's vast void of words. I prefer new words to old ones, Oriental or Occidental. The old ones are close to the soul, but even closer to rotting.

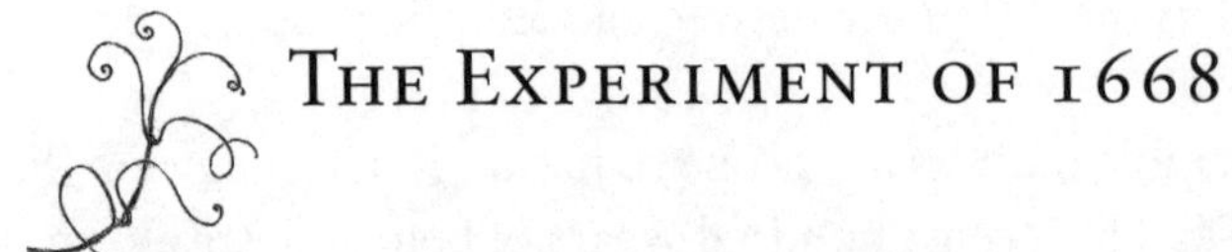

The Experiment of 1668

It's always astonishing to learn that things we take to be obvious have not always been known, but were once new and surprising discoveries. It's just as astonishing to consider what impossible and paradoxical things will be discovered next year or in a hundred years. But much of this sense of both the obvious and the impossible comes not so much from the nature of the discovery and of the world as from human nature: people hold on to their small certainties even at the price of great mistakes, their small successes even at the price of great destruction. In this sense, science goes against the grain of human nature. Let's consider a brief episode from the history of the problem of the genesis of life: Does life continue to arise from nonliving matter even now? And can an organism be generated from living matter even today?

These questions, if we begin counting from Aristotle, are a good twenty-three hundred years old. Insofar as there was a standard answer, it was broadly in the affirmative: fleas from filth, rats from rags and from slime, all kinds of insects from dew, flies from rotting meat, frogs and turtles from swamps in response to the beneficial effects of the sun, grass from clay, worms from mud, and eels from worms. Aristotle posited it, Pliny substantiated it. Christian church dogma was accordingly adjusted to fit it. It was more than a little worrisome, given the general creation of the world in six or seven days, that anything would go on creating itself, but the authoritative Saint Augustine accounted for the spontaneous creation of life as a kind of reverberation of the original creation. So it was all right.

Until the middle of the seventeenth century, it didn't occur to anybody to test the theory. Paracelsus gave instructions on how to produce a miniature person from urine and blood inside a gourd or in a horse's stomach. A natural scientist as renowned as van Helmont maintained that mice could be gotten from grain scraps and laundry filth. Alexander Ross, in a dispute with Sir Thomas Browne, who doubted the genesis of mice from rot, proclaimed:

> But we would then also have to doubt whether worms rise in cheese or in wood, or beetles and wasps in cow dung, or butterflies, grasshoppers, mussels, snails and eels and the like from rotting substances, which must always take the form to which they are designated by the creating force. To doubt this means to doubt reason, meanings, experience. Whoever denies this, let him go to Egypt, and there he will see a field settled by mice which rose from Nile's mud and will cause great disaster for the inhabitants.

That sounded convincing. Belief in self-generation was lodged firmly in people's heads, more strongly than many other beliefs.

But this was the dawning age of one of humanity's most powerful tools, the natural science experiment. In 1668, Francesco Redi, poet and court physician to a Tuscan grand duke, came up with an idea that was indeed uncommon. Francesco Redi tried an experiment in his chamber beneath the forested Arretine hills: an experiment so well conceived that its method precisely fit its needs, an experiment so simple that it required no implements other than those readily available to this distinguished member of the Accademia della Crusca. He had them on hand. Francesco Redi put on his table three pots containing raw meat. One he left open. Another he covered tightly with parchment. A third he covered with a small screen. Then he went off to write some sonnets.

In the days that followed, the inevitable occurred. The meat began to rot. Flies flew in and laid eggs on the uncovered meat. Maggots hatched and established themselves in their unique maggoty fashion. The flies could not reach the pot covered with the screen and had to lay their eggs on the screen. The pot covered with parchment they left alone.

Redi repeated his experiment, always with the same result. He also began to study the development of the insects. But from the original experiment, which was completely unambiguous, he derived this conclusion: life arises from nonlife only through the introduction of germs from outside.

Is that all? Yes, that is all. This immensely simple thing, the experience of every female servant buying meat on Friday for dinner on Sunday, was a milestone in the development of science. Similar experiments were held and defended by other scientists—Swammerdam and Vallisnieri, for example. And soon Redi could generalize Harvey's conclusion on the beginning of all creatures: all life begins in eggs. And it was generally accepted (against Aristotle's data, but in the spirit of his law of the unity of being). The self-generation of higher animals was forever driven out of the history of science, even though Redi himself still allowed for the spontaneous rise of internal parasites.

The question moved into the microscopic world. Leeuwenhoek discovered protists—called infusoria in his time—and bacteria. Needham and Buffon seemed to show that infusoria arise through self-generation. Spallanzani, Schultze, and Schwann defended the opposite view in subsequent centuries. None of their experiments, however, had the advantages of Redi's experiment: their methodology and technical possibilities did not suffice for the demands of their research, and their results were influenced by the random absence of germs in the air. In 1858 Pouchet again opened the question of the self-generation of simple organisms in hay infusions. Pasteur's authority ultimately decided in favor of the transfer of germs by air. But even in this century, the German pathologist Busse-Grawitz claimed to generate cells from acellular matter, not to mention the Russians who tried to generate bacteria from viruses and who, given time and party orders, no doubt would have generated another Busse-Grawitz from mitochondria.

There is still a lot of post-original creation in the popular mind and in the highly sophisticated minds of postmodernist philosophers, who reprimand science for being an end in itself. Indeed, science should not be the end in itself. That distinction belongs to Redi's pots and stinking raw meat.

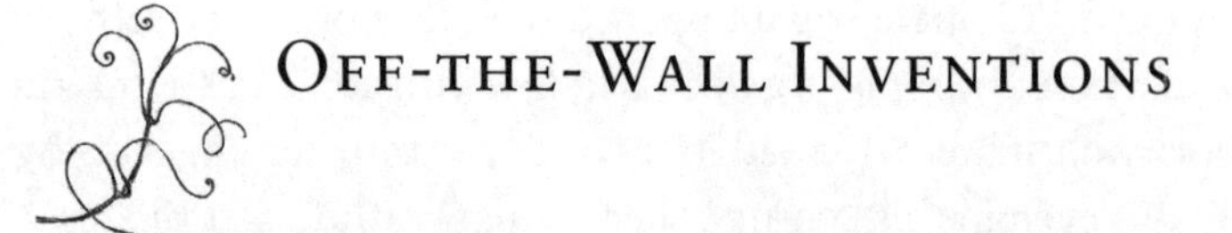

Off-the-Wall Inventions

The vast majority of patent applications are made by enthusiasts—that is, amateurs. The enthusiasm of professionals gets embedded in a tangle of technicalities and doubts, as well as in the gravitational forces of acquired routines. Inventors are people of the first contact, of unlimited optimism, of the splendid childish reductionism in which night and darkness result from the condensation of air and the evaporation of light.

In judging a patent application, I always worry that I am, basically, spoiling the fun. I recently had to turn down a patent application requiring that the human recipient of an organ transplant be subjected to a magnetic field in order for his cells, foolishly fighting the transplant, to be magnetized and thus rendered inefficient. Magnetized T4 and T8 lymphocytes are something very close to the condensation of air into darkness, but they are also very beautiful, like my first red toy car, which won the Monza race in our Pilsen living room.

Real patents granted in this field look rather tedious, with appellations like "Leukemia inhibitory factor from livestock species and the use thereof to enhance implantation and development of embryonic cells" (Australia Patent 622515 from 1992) or even "Novel fusion protein" (Europe 0 293 249 B 1, 1992).

The charm of the patents of those enthusiastic inventors is twofold: first, they tend to solve the really big and perennial problems of humankind, like Thomas Holmes's cadaver transfer bag and blood removal pump for embalming procedures. Second, they fill nonexistent needs, like John Dilks's solar-powered tombstone, featuring a video screen and the computerized voice of the deceased, prepared to converse with visitors.

My favorite patent is for equipment designed to eliminate train wrecks. At the front and back of each train is a car that slopes gradually down toward track level. Tracks are installed on the sloped cars and along the entire roof of the train. The tracks on the sloped ends of some cars are higher than on others. When two trains meet head-on, or when one train overtakes another, the train with the higher track setting rises up to the roof of the other train, rides along its length, and then drops down to continue its journey. Collisions are thus eliminated. This wonderful invention was patented on March 26, 1895 (United States Patent no. 536,360).

Fireproof suspenders were patented in 1885 (no. 323,416). Fastened to a normal suspender is a cord that uncoils far enough to reach the ground from whatever story of a building the trousers-wearer normally occupies. In the event of a fire and the absence of emergency exits, the wearer of the suspenders unwinds the cord and tosses it out the window to firefighters presumably waiting below with a cable that will be connected to the cord and pulled up to the window so that the endangered person can be lowered on it.

Patent number 174,162 from 1876 attacks a hardship less grave but more common. It offers persons who suffer unforeseeable attacks of hunger and hypoglycemia a sugarplum in some graceful shape (a heart, a flower), worn as a tasteful pin on the lapel. After the candy is eaten, the pin can be reused.

The electric bedbug exterminator (no. 616,049) came along

in 1898. This is a more complicated invention, demanding precise technical realization. Wires are fixed to a battery or some other source of electrical current, wound up the legs of the bed, and attached to connections on the frame and under the mattress. Distances are precisely calculated to correspond to average bedbug size. The bedbug proceeding up the bed will touch the wire in front of it with its front legs at the same moment that its back legs contact the wire behind it, producing an electrical shock that either electrocutes the bug or scares it to death. When this has happened to a sufficient number of bedbugs, word will presumably get around and they will stay away from that particular bed.

A patent for a rat and mouse exterminator (no. 883,611, from the year 1908) was inspired by similar psychological and humane concerns. The entry into a baited cage is equipped with a collar on a taut spring that withdraws the bait and circles the victim's throat. A tiny bell is affixed to the spring and the trespasser flees still wearing this equipment. The rodent then becomes, as the documentation tells us, a "bell-rat" that frightens its kin. When the number of bell-rats reaches a certain level, the ringing becomes unbearable for the others and the whole tribe flees the premises in terror.

The apex of technical skill may well be the automatic saluter, or hat-raiser. This is a very complex system of weights, pulleys, levers, and wheelworks hidden under the hat. The system is activated by a slight bow, which automatically raises the hat so that even citizens with their hands in their pockets or burdened with groceries can observe the social niceties.

Patents no. 949,414 (1910) and 748,384 (1903) provide for a person's first and last needs. The first is a teat at the end of a small rubber tube connected to a rubber bell of the appropriate format that tightly encircles the breast beneath the dress. Inserting the teat into the mouth of the infant induces a light pressure of suction on the bell and the breast, so that the life-giving fluid can spring forth without the nursing mother's being inappropriately exposed in public. The second patent provides for the preservation of the dear departed. The clothed body is dipped in liquid glass. After drying, it is cast in melted glass to the desired shape—the tetrahedron, cylinder, or pyramid, for example. If only the head is to be preserved, that can be similarly enclosed in a glass cube.

Such amusing patents may have inspired the pataphysics of Alfred Jarry, and subsequently the inventions of Marcel Duchamp and other dadaists. Those decades before and after the turn of the century seem a kind of pataphysically fruitful time that can no longer be repeated in this epoch of computers and microelectronics, which affect even the minds of tinkerers and handymen.

Despite the charming naïveté of such patents, one may feel a slight awkwardness in their presence. It's sort of like laughing at someone's limp. The one truly laughable thing is the office that registered them. A recent overview of patents by Brown and Jeffcott lists a patented trap for tapeworms (to be lowered by a thread into the empty stomach) and a set of metal plates to be attached to the temples for the purpose of extracting poisons.

Even more complicated feelings are aroused by the thought that in the last century, in the year 1822, Charles Babbage made the round of patent offices with his design for a differential machine that calculated functions from their measurable differences. In 1833, in cooperation with Ada, countess of Lovelace (daughter of the poet Byron), he proposed an "analytical machine" that had nearly all the properties of today's automatic computers. Babbage didn't succeed in convincing the chancellor of the exchequer of his machine's significance. With impassioned zeal he continued to refine it, and every improvement evoked the necessity of a further series of perfections, until he arrived back at the beginning and started all over again. Only Babbage saw these ever-minute steps of his as leading toward a final victory. The concerned officials saw only an interminable series of adjustments to an incomprehensible mechanism, accompanied by requests for financing something that seemed to them utterly useless—that is, in navigation or in ballistics, the areas that they considered practical. An automatic hat-raiser would probably have seemed less bizarre to them than an uncompleted automatic computer. Babbage died in 1871, five years before the invention of the lapel sugarplum, still solving substantial problems in the invention of the computer and still completing nothing.

With his work, Babbage contributed to the history of computers, which lack the myth of being one person's invention, like the lightning rod or the ship propeller. It's a field in which even inventions attributed to a single name are built on the work of others. Such a case is von Neumann's, whose conception of running two computational programs simultaneously without outside intervention is based on the forgotten designs of Mauchly and Eckert, who eventually lost their supporters and even their patent, which was taken away from them by a not very knowledgeable judge.

This history of discoveries and inventions is as splendid as it is tragicomic. It is difficult to expect of an office, however esteemed, that it separate the splendor from the tragicomedy. There can be no brain center to govern the nuances of the word *discovery.* It would have to be able to foresee the technical and developmental value of each invention, its entire technical and human potential.

But I can't rid myself of the feeling that the patented invention of a hygienic tin can for used chewing gum isn't something altogether imbecilic or contemptible. It can't be compared to the discovery of aspirin, just as the collision-proof train is not on a par with the invention of the steam engine. But it is the result of the play of the human mind, and such play, however laughable at times, is far better than serious indifference. To read a book about off-the-wall inventions is thus sprightly and dadaist on the one hand, and unsociably painful on the other. When you rise above it for an overview, it is also optimistic, like children at play.

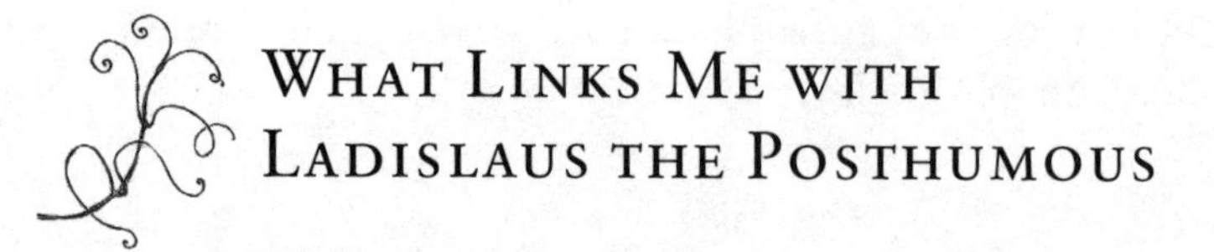

What Links Me with Ladislaus the Posthumous

I owe Ladislaus the Posthumous considerable gratitude. First, he was easy to remember in history classes: he's the only one of our sovereigns who was born four months after his father's death and who didn't do much of anything during his reign because he died at the age of seventeen, fifty-six days after his arrival in Prague. Second, it was thanks to him that I graduated from medical school.

He was a Hapsburg on his father Albrecht II's side, and a Luxembourg on his mother Elizabeth's side. He was born in 1440 in Komarno, crowned as a tiny infant and against the resistance of the Hungarian estates, with Saint Stephen's crown, and later made, without resistance, the sovereign of all Austrian lands. He was crowned with the Czech crown in Prague on October 28, 1453, after many skirmishes between the Rosenbergs and Lord George of Podebrady. George became, given the king's age, protector of the realm for six years. However, in 1457 Ladislaus returned to Prague to assert his claim to the throne and take a French princess as his queen. While he was in Hungary, he had managed to accomplish only a few dubious things: putting to death, rather perfidiously, the son of the local lord protector, Hunyady, provoking the uprising of the entire Hunyady clan, and possibly acquiring syphilis.

The Prague visit must have appealed to Ladislaus more as

recreation than as duty, because he jumped the gun by arriving in September rather than the scheduled November. It was on November 20, when he was leaving the castle, that "suddenly his head started to ache" and "the next day two tubers burst and he kept them secret because of the lower shame" (that is, they were located in his intimate parts, which he was unwilling to display to anyone). Shortly after, he presided at a trial and then sat down to a five-hour banquet at which he appeared "very cramped." Afterwards, he had beer and radishes brought to his bed, which cheered him up. However, after lying down he got terrible pains in the stomach area. Toward dawn, doctors were summoned and came up with opposed diagnoses ("King, nothing is wrong" and "King, it goes badly with you"). Their remedies included thorough bloodletting, to get all the poisons out of the body, and resulted in vomiting and sweating. The seventeen-year-old Ladislaus survived all that, but the next morning he could not rise from bed, but lay crushed and comatose, and after another bad night, while he was "lying in a faint, this handsome young lad's soul abandoned him on Wednesday of Saint Clement in the hour of XXIII." Around his eyes and nose and on his tongue he had petechiae, little blood spots, and larger dark spots appeared on his chest and abdomen.

"And before King Ladislaus died he asked Lord George, the Protector, and this in sound mind and memory, to accept his Czech country and rule it and help orphans and widows to obtain justice." This must be the lyrical license of the Czech chronicler, for there are no other reports of Ladislaus being more compassionate with the Czech territories, including the orphans, widows, and ruling protectors, than he had been with Hungary and the Hunyadis.

In any case, according to a physician's testimony, after his death another "tuber burst out" and, according to a certain Peter Eschenloer, the king's belly was conspicuously swollen, even though it was cold, and so the burial was accelerated, taking place within two days of his demise. Even the customary halts of the funeral procession at individual churches were omitted. The body was exhibited in the Royal Court in the Old City, but the barber-surgeons, frightened of the black spots, refused to embalm it, fearing black plague.

The official report on the king's death from Lord George, addressed to the Hungarian estates on the day of his demise and to the Vienna estates five days later, speaks of death by natural causes. A dispatch from the Venetian ambassador to the Prague court states that cause as "Belgrade plague," while the report of the Czech ambassadors to the Austrian estates in January 1458 plus an opinion by an entirely neutral Jean Naupoint describe the cause of death as an infection.

Reports are reports, descriptions are descriptions, and interests are interests. In the nineteenth century inflated nationalist passions latched on to the golden-haired king who was exhumed from the tomb and reburied ten times or so, without further medical diagnosis or concrete evidence about him except for the fact that his hair must have been very ample and very gold. The romanticizing German and Polish historians, in accord with their pious intentions, have Ladislaus being killed by arsenic, by the hand of Lord George, naturally. Equally enthusiastic Czech historians, endorsed by the authority of the founding father of Czech history, Palacký, have Ladislaus killed by everything from plague, syphilis, botulism, ulcus molle, and other sexually transmitted diseases to Werlhof's disease from lack of blood platelets, salmonellosis, and foudroyant sepsis.

Romanticizing history is like chanting spells and performing rain dances in an autopsy room. The central theme or intention functions as the nucleus that over time gets wrapped by an accretion of woolly verbal material, somewhat the way cotton candy is produced at a county fair. If you want a poisoning, you specify adequate doses. If you want natural death, you surmise various refined forms of infection and, insofar as your education or adviser allows, metabolic disorders and endocrine anomalies.

As it happened, I had to sit all the exams of my second and third medical rigorosum (a set of six exams, usually before a committee), all twelve of them within a period of four months. Four hundred ninety-six years had passed since the death of Ladislaus the Posthumous. At that time, however, I happened to read Gustav Gellner's condensed account summarizing all the available historical, pathohistorical, and pathographic information on King Ladislaus, along with the various speculations; some parts stuck in my head.

The first in my heroic succession of exams was pharmacology. By some chance it was not the pharmacology professor who showed up, but the psychiatrist and psychologist, Professor Vondracek, a great and rather liberal personality, original both in his appearance and in his thinking. He had me prescribe some barbiturates, which did not cheer me or him in the least, and then he crossed over to trivalent and pentavalent arsene as the basis for treatment of syphilis. When I was listing (or trying to list) the symptoms of mapharsene overdosage, it occurred to me, aloud, that they somewhat resembled the premortal difficulties of Ladislaus the Posthumous. Though of course we, together with the great Palacký and as good Czechs, I added, don't admit to any poisoning . . .

Professor Vondracek suddenly became very animated. He announced that he too favored an infectious cause of death and as a good Czech condemned the conclusions of the evil Germans Kanter and Lewin. A deep inner sympathy burst out between examiner and examinee, and we were on the verge of patting each other's shoulders. Vondracek promptly assumed that a student who could so organically relate medicine to the life and welfare of the nation should not be bothered with some prescriptions and specifics of diagnosis. We enjoyed such a prolonged and amiable conversation that the medical students waiting in the next room took on the color of limestone and asked, when I finally emerged, "What did he torment you with so long?"

"He didn't torment me," I replied, "we only spoke about Ladislaus the Posthumous."

"Jesus," exclaimed the man who was next in line to be examined, "that's not in the textbook!"

Of course, I pulled an A in pharmacology.

Not long thereafter I had to cope with pathology, which used to be the exam of exams. I smoothly switched from endotoxin shock to the postmortal protrusion of the royal stomach and quoted the pertinent testimonies of the Czech historians. Professor Viklicky, too, was impressed with my broad historical and pathological views, and in spite of some dubious aspects in my nephrosis/nephritis concept, he invited me to join his staff, upon graduation of course, which he thought would be soon and with success.

Leaving the pathology department of Bulovka Hospital that day, I realized that for me the heart of the matter lay in Ladislaus the Posthumous. Not as a supplement or as the basis, but as what the Marxists call *survalue,* the profit resulting from exploiting workers.

So it was that with Associate Professor Skalickova, who ambushed me with a question about Pavlovian interpretation of schizophrenia as a spread-out suppression, I successfully got out of the trap by discussing progressive paralysis with adjoined infection, which could be easily accommodated with the historical account of how Ladislaus passed November 22 and 23, 1457. Meanwhile, the professor was grabbed by a paranoid schizophrenic confined to a barred bed; he caught her hand through the bars and would not let her loose, which was rather embarrassing for an examiner, so that she obviously took my historical excursion as evidence of high social tact in the examinee.

Thus it happened that in dermatovenereology I attached to condylomata lata, which was actually the patient's malady in my exam, the possibility of a malignant syphilis, or of panniculitis, as the cause of Ladislaus's recorded "blisters." Professor Gawalowski, as an enlightened and educated man of the old school, thus forgave or overlooked my low level of knowledge of atopic eczemas.

I was even able to lift the spirits of the old pathophysiologist, Professor Hepner, who was deeply depressed by the "circles" of medical students who sat collectively for the exam and whose collective wisdom sometimes seemed even duller than that of any given individual. I did it by an epic description of petechiae, the blood effusions that had been noticed on Ladislaus's skin. Since Professor Hepner had just failed a student who stated that vitamin C was a white fluid that, when given to a bird, killed it, my Ladislavian excursion was taken as a gift from heaven, and puffing on his cigar in a stately way he gave me a mark that only Werlhof would have deserved.

During the exam on infectious disease by Professor Prochazka, I slipped in Ladislaus's paratyphoid, or an eventual botulism, while in pediatrics I triggered a little respect from Professor Houstek by exploring Ladislaus's possible juvenile diabetes, thereby diverting attention from my ignorance of Mohr's reflex.

I grew so confident about the detour to Ladislaus the Posthumous that I started allotting to the poor King Lackbeard other likely disorders, anticipating a literary genre that flourishes today. Thus, in the internal medicine exam under Associate Professor Fojt, while marking the position of the liver and lungs on the surface of a good-natured and cooperative patient, I playfully switched to hepatitis and terminal pneumonia in you-know-who. In neurology I sketched his possible multiple nerve inflammations from infectious or toxic causes. And in the ear, nose, and throat exam I laid before Associate Professor Chvojka the eventuality of infectious mononucleosis with a general weakening of the organism by an abortive throat form of bubonic plague. Professor Chvojka seemed to be really pleased, but my grade sheet got lost in the dean's office (most human endeavors get lost in an office), so that I had to come back to the clinic and—to prevent a new exam—draw Chvojka's attention to the fact that I had been examined already and that he might remember it by my discourse on Ladislaus the Posthumous. Chvojka, a man of an aristocratic noblesse, politely "remembered."

The only exams in which I didn't succeed in introducing Ladislaus the Posthumous were gynecology, in which I got a B, and surgery, where Professor Knobloch insisted on my estimating the invalidism of the patient before us and rejected all references to the invalidism of King Ladislaus, principally on the ground that kings are not itemized in medical bills.

After I completed my medical studies, I didn't need poor Ladislaus anymore, since in the freshly developing field of immunology it was more important to deal with the excesses of Soviet biology than with the long-ago difficulties of Czech kings.

Nevertheless, I hold Ladislaus in grateful memory, and have kept up a personal interest in the results of the "medical and anthropological investigation of the historic personalities in Czech history." I was especially pleased to learn recently of the results of Dr. Emanuel Vlcek's investigation of Ladislaus: "Skeletal remains were permeated by pathological foci literally from head to toes . . ." These changes were compatible with the conclusion that Ladislaus the Posthumous was bumped off by a tumor of blood-forming tissue in the form of acute lymphatic leukemia, naturally with an infection as the terminal complication.

It's worth noting that all the evidence under consideration, both in my medical curriculum and in the Czech, German, and other historical literature, was just pious speculation, in the face of which reality is frightened off.

When a sensitive amount of crucial information is missing, our capacity for fantasy and association takes over. Even if it is endorsed by specific terminology, it is just a singing exercise in one key and one octave only. Exercise is a good thing, and may further the evolution of the individual, as demonstrated here, but nothing more.

For me, Ladislaus the Posthumous doesn't merely have the value of an auxiliary examination tool. Ladislaus the Posthumous is a striking example of the theorem that states: How come you didn't think of leukemia, you blockheads?

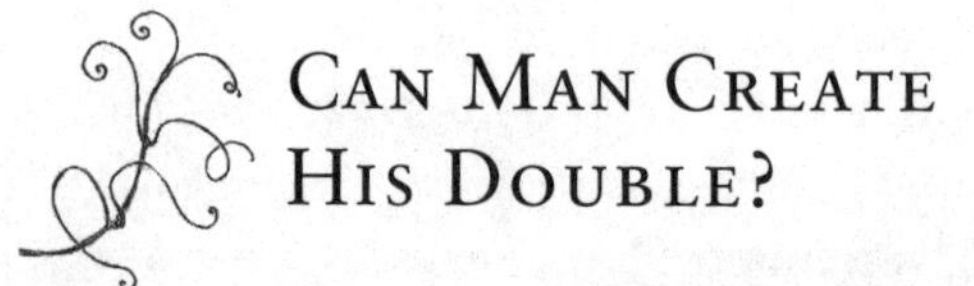

Can Man Create His Double?

The question in my mind is not so much whether he can, but why he would want to.

For my part, I never wanted a double. I can even say that I would hate to have one. Not because I feel unique, the most important phenomenon since the Big Bang, which seems to be an acceptable view for thoroughbred poets to take of themselves, but because I would hate to have all my errors repeated, all those committed in the past and all those yet to come. I used to dream, instead, about being invisible: a way to escape enemies, catch criminals, inspect the teacher's notebook to find out about the next math test, and, most of all, to see what my girlfriend was up to when I wasn't around. I would especially hate having a double visiting somebody I really loved.

I had another daydream that I am afraid was not that irrational, and still isn't. I imagined that most of the important politicians must have a double, or better still, a team of doubles. Not only to cheat the terrorist, the enemy, or the missile, but also to assure political stability. It makes no sense that a simple clogged coronary artery or senile dementia should affect the fate of the state, the party, the battle—usually victorious—or the campaign, say, to introduce penguins into the Aral Sea area to cover up the fact that it is disappearing.

I speculated that in the case of an untimely disease or death, the first double in line, having learned all the gestures and the vocabulary—not too difficult in most cases—could make the public appearances. Of course, if the great person in question had a family, the double would have to fill the roles of father, grandfather, and husband, but that wouldn't be too difficult either, since great persons are usually pretty neurotic and a well-trained double would be much easier to live with. If the original statesman happened to survive, he would have to be shot, or maybe disguised as a somewhat atypical penguin. When the first double wore out, he would be disposed of and the next one would appear, and so on until the moment when the death of the serialized statesman became admissible or useful. In prolonged serialization, of course, even the family might have to be duplicated in this way, and the whole operation would require a really good secret police service, especially well-read in Freudian literature. Without a good secret police, one would have to double the capital, even the whole nation.

Actually, this duplication was attempted in a rather primitive fashion by the Nazis and in communist countries where ideal citizens held identical opinions, and, through the art of socialist realism, sported identical rosy complexions. Karel Capek, the Czech writer who would have won the Nobel Prize if he'd been English or Italian, built an impressive metaphor of redoubled, identical, possibly cloned beings in his novel, *The War with the Newts,* first published in Czech in 1936. It was one of the earliest catastrophic science fiction novels, and it ended with the newts taking over the human world and appearing finally in the heart of Prague, in the Vltava river. In this case, science fiction wasn't too distant from the realities that were to follow in Central Europe. But now, at the demise of the communist psychopathological experiment in cloning human minds, we can state with a grain of biological satisfaction: Hot damn, it didn't work! We were even more diversified at the end of the experiment than at its start.

Creating a double, a replica—in other words, cloning a human being—is, as to biological reality, still deep in the realm of fiction. The essential virtue of science fiction lies in the fact that science is poured like a Thousand Island dressing over the fiction salad.

In the 1970s, experiments with the transfer of a nucleus from an intestinal epithelial cell of the adult donor into the zygote could be repeated in frogs, but not in mice. You could implant the entire genetic outfit of a cell into a fertilized ovum (zygote) of the given species and, upon incubation of the cheated egg and transfer of it into a foster mother, you could get an exact replica of the donor of the cell. With more cells and more zygotes you might get a full house of replicas: beautiful, shiny, uniform clones of the donor. Because the donor could be an adult organism with known qualities and capacities, you might have a rational choice of what, or even whom, to clone.

As is usual in science, the failures involving the irreproducibility of the transfer of an intestinal cell into the ovum did not stop further attempts to solve the soluble, and now we have cloned sheep and monkeys. It turns out that not even the high sensitivity of mammalian cells to the difference between the gut and the ovum, and to the position of the donor cell in the cellular cycle, is a limit. The resulting cloned sheep and monkeys are not horrors of the brave new world any more than human identical twins are. Human identical twins occur spontaneously and are considerably influenced by the environment, which affects their intelligence and even the onset of mental and physical diseases; they don't even have identical fingerprints. A kind of cloning occurs as a regular feature in some species, like the armadillo, where the developing embryo splits in the four-cell stage into two individuals of the same sex, almost identical (with a possible effect of mutations). No journalist or philosopher was ever alarmed by armadillos. Natural cloning appears to be less objectionable than artificial, human-instigated cloning, which will still be, for some time to come, more successful in the lower vertebrates.

This is a general experimental and medical difficulty. Increased complexity of the body's internal organization and increased sophistication of the immune, endocrine, and nervous systems bring with them a considerable narrowing of the possibility for manipulation of the body and of its own intrinsic ability to forget injuries and to reassemble and regenerate.

A sophisticated internal organization, the fact that mammals, ourselves included, are filled not with a simple trickle of the music of life but by the scherzo of an extremely richly orchestrated symphony, does not predict anything about the prospect of survival on the planet. With regard to ionizing radiation, for example, any crab is more apt to survive large doses than any field marshal. The field marshal only has the advantage of more information on the timing and topography of the event and more choices about how to get away.

The earthworm can regenerate a considerable part of its body. A newt can regenerate an entire limb. We can't regenerate damaged or lost organs; we can only accept transplants, and this only after three thousand years of trials and errors in medicine. To invoke the earlier metaphor: only after we have learned how to partially deafen the conductor of the inner symphony. It doesn't say anything about our survival, only that individuals can survive their hearts, pancreases, kidneys, lungs, and livers.

Hypothetically, clones of people with a special resistance to dominant diseases might guarantee longer survival in a given moment. But in the next moment they would be in trouble, since the diseases and their agents would change and there would be nobody to be selected for coping with the new situation. We would have to keep unchanged even the external agents, viruses and microbes, and ensure a stable environment, ranging from radiation and temperature to acidity of the waters. We would have to bring the world to a standstill, to make it a paradise of replicas and nondiversity.

This is the same story as that told by Lewis Thomas: Suppose you wanted to clone a prominent, spectacularly successful diplomat. You would have to catch him and persuade him. You'd have to be sure to recreate his environment, perhaps down to the last detail. So, to start with, you would probably need to clone his parents and, for consistency, their parents as well. This means the diplomat is out, since you couldn't have gotten cells from them at the time he was recognizable as a potential social treasure. But this is only the beginning. It is the whole family that really influences the way a person turns out. Clone the family. But each member of the family has already been determined by the environment, schoolmates, acquaintances, lovers, enemies, car pool partners, even peculiar strangers across the aisle in the subway. Find them, and clone them. But each of the outer contacts has his or her own surrounding family and their outer contacts. Clone them all. You must clone the world, no less. It would mean replicating today's world with an entirely identical world to follow immediately, and this means no new, natural, spontaneous, random, chancy children. It would be our world, filled to overflowing with duplicates of today's people and their duplicated problems, probably all resentful at having to go through our whole thing all over again.

"I once wondered," Lewis Thomas concludes, "what Hell would be like, and stretched my imagination to try to think of a perpetual sort of damnation. I have to confess, I never thought of anything as bad as this."

So the story of creating doubles appears not to be a very optimistic one, not from the biological point of view, nor from the psychological or sociological one, statesmen aside.

Nature has, of course, attempted doubles. They are

successful in plants that propagate by runners and by budding, which we extend by graftings and cuttings, and even tissue cultures, which yield a new, but identical, carrot out of one single cell of the emeritus carrot.

Cnidarian polyps reproduce asexually by budding and forming colonies when they're attached to each other, like the mythological Geryon (whose prototype was the strongest man alive, with three bodies and six arms). The polyps are genetically identical; they acquire quite differentiated functions and often become mere organs in the assembly called a siphonophore: a disgraceful fate for an individual. Maybe doubles are better off as nonindividualized units, which is the rule at the cellular level, even within our bodies. Clonal expansion of identical cells is the basis of our immunological defense.

Less typically, some aphids' mothers produce their own replicas by laying unfertilized but merrily developing eggs, as do some rotifers and ants. (Cloned aphid replicas have even been described as mere parts of a single motile superbody.) To keep the aphid world on the go, this generation of doubles must be succeeded by a sexually produced generation; there are some exceptions, like the European woolly apple aphid, which misses its host for the alternate sexual generation, the American elm. But this is a human-induced exception and may not prove successful in the long run, like other imports.

It appears that the production of doubles is abandoned as life proceeds toward internal complexity and sophistication. Doubles, replicas, and clones can function only as subservient units of a lower order, not as real individuals.

On the human level, only simple minds dream of doubling themselves. Or some unsuccessful ones, hoping for correction of all their misdeeds in the next life.

But a carbon copy can hardly be considered a next life.

Apes, in Particular

The special attention that animal rights people accord to apes and monkeys is only another kind of Christian anthropocentrism. Monkeys have those little faces, almost like ours, they have little hands like ours, and they have eyes like ours, so no doubt behind their eyes they also have souls. We pity monkeys as almost-people while we don't really care much about the capybara and the pangolin because they look more like capybaras and pangolins than they look like us.

Yes, the eyes of apes are sad and black and reflect what Durs Grünbein noticed:

> that sadness of not being born to anything
> but just an animal,
> that evil futility, accepted
> with a motionless face.

Alas, the basis of animal protection and compassion is only human solidarity with human similitude, not compassion with anything that's writhing in pain, for pain is dictated by the pitiless laws of natural selection or primitive human rituals.

It is the immeasurable good luck of *Australopithecus* that, thanks to the operation of those pitiless laws, they didn't live to see our enlightened age. If they had, they no doubt would have become the subject of the same legal ponderings and radical street solutions as a child-to-be in the form of a cleaving egg, which differentiates progressively from a bundle of cells to organs to a fetus that, on top of that, opens its little mouth in a so-called silent scream. Is *Australopithecus* a human being because he walked on two legs and opened his mouth or because he had an immortal soul? If he didn't have it, would he have acquired it, had someone existed at the time to christen him? Anoint him? Educate him about australopithecine rights?

Apes and monkeys had very bad evolutionary luck in that they survived alongside the human branch. As a result, certain people, close to nature, a.k.a. primitive, eat monkey testicles and smear themselves with monkey blood infected with the HIV virus while educated types, removed from nature, lament the suffering of monkeys and shut them up in cages so they can lament even more over the apes' and monkeys' laments, which are, probably, something completely different from what people think.

The bad luck is that much worse because only monkeys, given their evolutionary close kinship, are suitable for the testing of certain substances that are going to be used on people. So, if we can protest somewhat sensibly against shutting apes up in cages, we can't protest very much when it's a matter of an alternative to experimenting on people.

The indignation about cages becomes dubious compared to the plight of children shut up in apartments with psychopathic parents, incestuous or sadistic fathers, alcoholic mothers, or bigoted families who punish disobedience with blows and treat leukemia with prayer.

The specific ape question in our case was: do we test an antibody that can control undesirable reactions to transplants in people (whose lives are at stake) on primates first, as the legal regulations require, or do we protect the rights of primates and light a candle or call on Allah for the success of the transplant?

With regard to religion, it should be added that the situation of parents facing the necessity of a transplant for their child is closer to a kind of slow crucifixion than to a sojourn in a cage.

The research team went to the zoo, where a pair of gorillas was confined; the male wasn't well anymore anyway, and the female gorilla suffered from an invisible arteriosclerosis and a visible eczema, which the experimental substance might even help her with.

The veterinarians put the lady gorilla to sleep from a distance with a tranquilizing dart and then let the medical team in. The team stepped into the cage nervously, because even dissolute human patients have options and manners in comparison with gorillas. The husband gorilla, in the neighboring cage, ascribed very low motives to the people in white and commenced defending his lady by the only means available to him: he skillfully hurled excrement at the white coats and roared like a bull. After a direct hit on his good jacket, the local veterinarian retreated behind the main door and left the human doctors to their fate; they, after all, were wearing public-property outfits. When the male had exhausted his ammunition, the doctors managed to get in to the female, who was snoring in the corner like a big, furry black log.

They managed to anesthetize her and inject the appropriate dose of the antibody, and all they had to do after that was to take ten-minute blood samples, from which the effect of the drug and its action would be evident. After the first ten minutes, the blood drawing was successful. After the second ten minutes, they took out a sterile test tube and unwrapped a sterile syringe and needle and turned back to the cage and its corner, to the accompaniment of the male's continuing roars.

However, there was no log in the corner. The female gorilla towered there, virtually larger than life size, and she began to march between the frozen researchers like a golem from Jewish folklore, toward the cage's open door, with fateful paces, step by step, until, with a curiously smooth motion, she jumped down from her cage and made her way along the aisle to her husband's cage. The only move available to the researchers was to close the cage they were in and huddle down inside it. The male gorilla's roars diminished, leading the sullied veterinarian waiting outside the ape pavilion to assume that things were going smoothly.

Thus it was that this unheard of situation arose: the researchers in the cage and the experimental animal on its way to freedom. The researchers in the cage couldn't change their situation because they were afraid even to let out a peep. The gorilla was in a position of free choice, limited only by geography and by the time it would take somebody on the outside to catch on.

It would have been a well-deserved and symmetrical situation if the gorilla had been able to use it to research something that would help gorillas. Unfortunately, she couldn't. The planet of the apes, according to the popular film, can hardly happen here. The trouble with science fiction is not excess of fantasy, but not enough facets to the fantasy.

In a certain idealistic sense, apes' rights were upheld. Alas, apes have a hard time understanding the idealistic sense, so a certain adjustment of their rights, as we understand them, is really of no use to them.

However we consider it, we can't reverse the basic human-animal situation. The standoff with the gorilla was solved by another tranquilizer dart. It was, given this world, a humanitarian solution. It would be fortunate if it were possible to solve, in an equally humanitarian way, the situation of people whose kidneys fail, and if it were possible to solve the situation of children whose parents fail them.

If, at the social level, we are going to have experimental

children, I don't see a practical solution to the problem of experimental animals, other than the exercise of basic human feelings, which are not incompatible with research but, in fact, just the opposite.

On Pigs As a Laboratory Vacuum

THE LABORATORY VACUUM is produced neither by sucking out air nor by vacuum-cleaning. The laboratory vacuum is a permanent phenomenon, based on the circumstance that something can be too small for the human eye to make head or tail of, too tiny to be manipulated—that is, physically handled—and at the same time too big and complicated to be viewed under the microscope or analyzed biochemically. The subvisible is no problem; there are micromethods for viewing it. The gigantic is even better: you can get spectrophotometers and radio telescopes to measure it. But something else, by comparison, may measure an impudent 0.4 millimeter, and it can't be turned into homogeneous mush and it can't be manipulated by a micromanipulator, which is a laboratory robot made for microscopic movements induced by a thumb and forefinger. That predicament induces a mental stress compared to which Artaud's theater of cruelty is child's play.

Of course, there are many who have developed techniques for operating in the laboratory vacuum, or so we learn from their communications. In my field I regard it as an incredible laboratory wonder that someone could remove an organ of the category of the omentum from a thirteen-day-old mouse embryo, which is itself the size of a gelatinous tear shed by a dwarf of the smallest type.

As a matter of fact, organisms are improperly organized for scientific technologies. In experimental biology it's only the rabbit that, by its size, proliferation, and easy manipulation, seems to have been created appropriately, and what a joy it would be to have rabbit-sized mice on the one hand and mouse-sized cattle and pigs on the other. That's because horses, cows, and pigs are more suitable than rodents for studying the beginnings of immune response during the development of the fetus and after birth; rodents in the neonatal stage are defended by their mother by way of immunoglobulins, sometimes even cells, passing through the placenta or yolk sac. Horses, cows, and pigs have an impermeable placenta, and if they are prevented from suckling colostrum and, later, mother's milk, they develop their antibody responses from scratch and independently. Since these responses are influenced by microbial colonization, it is desirable, in terms of research, to keep such models in a germ-free state, in an incubator if possible. It stands to reason that a calf in an incubator is not very common, but there are breeds of dwarf pigs, and their sucklings are substantially more practical both by size and in terms of their level of general mess-making.

While American researchers could make use of the so-called Minnesota white "minipig," we Czechs, with great pride—that is, limited financial resources—developed in the early sixties a Vietnamese socialist minipig. It was supplied somehow from Vietnam, then at war, to the academic farm named Chvatalka, south of Prague, not far from the famous René Hotel, which I imagined, while I was at Chvatalka, to be something like the seat of the Holy Grail, with all the Round Table types in attendance, and many beautiful ladies whose silvery laughter drowned out the hoarse grunting of the academic pigs beyond the woods.

In order to catch up with bourgeois pseudoscience and surpass it, we had the Vietnamese pigs fed and groomed with exemplary care. They were rather disgusting creatures with wrinkled, depressed snouts, and they were dark gray and on the bony side. They resembled a semibald porcupine caught in a frontal collision between two armored cars. The creature's tail was rather like a withered willow branch. These pigs did not come in white; the Vietnamese sun permits only pigmented pigs.

The development of the Vietnamese minipigs to the reproductive stage was monitored carefully. The blackish minipigs wallowed in their muddy pen, made sucking noises when they were feeding, and got bigger every day. A year and a day later we had a penful of monstrous minipigs, the largest in the world. Since, as veterinarians say, they were "basically small-framed," their growth at Chvatalka resulted in something with dachshund feet and a Saint Bernard belly. The belly dragged in the mud wherever the pig went. Their faces, meanwhile, were no longer depressed in the Vietnamese way; they were more like those of melancholic Czechs who cannot squeeze into their own pants. Their amorous frolicking was reminiscent of a sea slug being squashed by a heavyweight mud god, and it called up associations with succubi and incubi and the devil and Rosemary's baby.

In terms of zoology, the Chvatalka farm managed to complete the domestication of the Vietnamese minipig, a historical victory of matter over matter; sows in our country and elsewhere in the West weigh up to 80 kilograms, and male pigs reach 120 kilograms.

In terms of the butcher's craft, the pigs run to fat early. That may be a genius loci, considering local human types: Czech black market specialists or youthful entrepreneurs, inspired equally by foreign supplies of dirty money and by pure genome.

The tender meat of Vietnamese minipigs, says a butcher of any type, matures later.

In terms of laboratory research, however, a small error must have crept in, because the resulting pig, by its format and its tendency to mess and stench, surpassed even a scientist's gloomiest dreams of a dinosaur submerged in Jurassic scrunge.

Monitoring of newborn Vietnamese piglets, which were never suckled and would therefore develop an independent antibody repertoire, never occurred. Nevertheless, the world's biggest minipigs went right on reproducing, and as a consequence a working team was set up to eliminate the maturing individuals by means of Chvatalka pork feasts, obligatory even for research students. With the participation of a specialized butcher and scientific kitchen volunteers, not to mention crates of beer and clinking wine bottles, the world's biggest minipigs with late-maturing tender meat, having failed in the spiritual evolution, were used to develop the body mass of scientific workers, unless they managed to escape to the René Hotel.

I'm not sure how this would be handled by bookkeepers in a grant system, but it would probably have an impressive name, such as "Metabolic Methods of Eliminating Potential Sources of Immunoglobulins Produced without the Effects of Colostrum."

Impressive names aside, this was an interesting transfer of an animal—the pig—from the category of laboratory animals, wept over by kindhearted humanists, to the category of slaughterhouse animals, not mourned by very many people because the meat-processing industries wouldn't give a damn anyway, to put it politely. Vietnamese minipigs moved from the mathematical set of approximately 20 million mammals used annually in world laboratories to the set of over 5 billion mammals slaughtered annually in order to make lunches and dinners, and also breakfasts, in the United States alone. Of course someone may also adopt a Vietnamese pig as a pet (this has become something of a fad in the United States, where their potbellies are apparently considered very cute), like a dog or cat, moving the piggy to the set of 300 million pets, which rank among the mammals believed by humanists to have a great life—never mind that it's a life comparable to that of paranoid patients locked up in psychiatric wards, or slaves on Santa María, or eunuchs in an Ottoman harem.

Brehm records the example of a "Chinese pig," trained by his master, a forester, so that it followed him, like a good dog, came when called by name, went upstairs and down, never behaved improperly in the living room, hunted morels in the woods with great passion, and learned many engaging tricks, like that herd of piglets that danced to bagpipes to cheer up a glum Louis XI. Every decent humanist believes that a pig treated like that has all the fun we've been waiting for since the Neogene era.

Bypassing this psychological domestication, the Vietnamese minipig evolved from the status of helpless laboratory subject to the status of sausage.

The Vietnamese minipig, in my opinion at least, extends the notion of the laboratory vacuum to the socialist plans of scientific development, which usually conclude with reports in an appealing verbal package intended for the silken embrace of an official whose grasp of science covers Pliny the Elder and sausages.

It must be added, however, that the Vietnamese minipig has been repeatedly used in research facilities to revive and improve the breeds of the Minnesota minipig, which has in the meantime been imported to replace its overgrown Vietnamese ideological porcine counterpart.

That's how the matter got into the area of the early running to fat, the genius loci, wherein the entrepreneurial import is enhanced by an admixture of merry socialist features, such as the capacity to cheat anybody, anytime. Nothing swinish is alien to bureaucrats (the term *swinish* being used not zoologically, but ideologically).

This is the genius loci that can also be found in the laboratory vacuum we might call "sociological." It shows a tendency for universal mess-making, and it cannot be manipulated, not even by a micromanipulator. Nevertheless, it is the right stuff for Artaud's theater of cruelty, and humanists can just go jump in the lake.

The Electron Microscope Tesla 242 D

IT'S ODD that most of the major scientific discoveries anyone has ever made, in our country or elsewhere (or, more soberly, all the semidiscoveries, quarter-discoveries, and other fractions of discoveries), were typically achieved under immeasurably uncomfortable, virtually neolithic conditions.

The researcher was accommodated in a former warehouse repainted in scientific style. The microscope was a Zeiss, but from Jena. The centrifuge was Polish. The electrophoresis was homemade. The ventilation was out of order. A colostrum-free piglet fell into the pipe shaft. The power went out regularly. And the ceiling leaked. Even the leg of a stool broke, which I recall with special interest, because it wasn't under me and the sitter in question broke his hand as a result, which initiated a rather complicated inquiry, necessitating lots of official poetry, into whether it was a work-related injury or not. I think in the end it was concluded that it was a work-related injury but, in the context of the revolutionary trade union movement, that led to the same result as if it had not been—that is, no reimbursement.

As a result of discoveries made, even under the auspices of the revolutionary trade union movement, the researcher, together with his co-workers, sometimes hoisted himself up a rung. He was given sturdier stools, chemicals from Miles and

Gurr, appliances from Zeiss Wetzlar, a laboratory on the second floor, and a personal thinking room with an armchair or sometimes even a couch.

However, in this phase of scientific development not many discoveries, or their fractions, were made. This change may also be a function of age, since we work under neolithic conditions when we are young, but under conditions of relative, sometimes even bourgeois, satisfaction at a ripe, or overripe, age.

From this perspective I consider it unusually fortunate that during the era of communist "normalization" I was cast out into a shed that had originally been a breeding facility for small and midsized laboratory animals, and one where the ceiling leaked.

Having been cast out, I began to research with neolithic diligence. My subject was the thymus of thymusless mice. It wasn't until several years later that higher-ups began to inquire about what I was actually doing, since a thymusless mouse officially and logically has no thymus. I defended myself, referring on the one hand to the general law of science that everything is one way but simultaneously also the other way, and on the other hand to findings I had made.

These findings were made, quite stylishly, on, among other things, a simple—perhaps the simplest—electron microscope, the Tesla 242 D. This microscope resided in the subterranean chamber of another shed, probably a former warehouse, where water didn't just leak but occasionally flowed in a stream. The LKB ultramicrotomes didn't handle it well, while the Czech-Moravian microscope Tesla 242 D didn't mind at all. It was the smallest electron microscope in Europe, maybe in the world. But it made good enlargements, up to 45,000 times, which is quite sufficient for cells, and it worked reliably when it was handled by a reliable electron microscopist. Dr.

Rossman is, and was even then, an extraordinarily reliable microscopist. He focused with an unerring eye, estimated exposure times with a characteristic whisper of oneandtwenty-twoandtwenty-threeandtwenty, and stabilized the specimen holder, which tended to move of its own will, with a match. He developed negatives immediately, so there was prolonged darkness, as in a prehistoric cave, but even in the dark it dawned on us what had been captured, and how. A little, a little more, then a grid expertly called a "baddie" or a "cantie" (from "can't be done better"), then a "goodie" or an "almost decent one," and, yes, this one, this could be it. Electron microscopy would be a challenge to Cinderella, who'd have to cope without help from little birds, without even the puniest prince, and in absolute darkness. But she'd have to sort the lentils anyway.

Electron microscopy is best understood through the story of the Tesla 242 D itself. This microscope, by a turn of fate unknown to me, appeared in the Rockefeller Institute in New York in 1965. It attracted a certain amount of attention because products from our geographic latitudes have never been known for either their miniature size or their reliability. The New York 242 D worked, and worked surprisingly well. The microscopist was as proud of it as Prince Charles might be of slaying an eight-point buck with a small bow and arrow.

The time came to do maintenance on the microscope. I don't know why, but in America the maintenance of electron microscopes is frequent and obligatory. The microscopist took the 242 D apart and laid the components out on the table. For such a small microscope it took a very big table. The view from the open door was imposing: there were rods, screws, boxes, pins, wires, cylinders, holders, apertures, diaphragms, filaments, covers, clamps, and mirrors from noon till night, from north to south.

But then at lunchtime, when the microscopist stepped out, some joker came by. America is full of jokers. That's why we feel at home there. The joker noticed the imposing acres of components, hit on a joker's idea, and ran off to the supplies department, where he got a chrome-plated screw and deposited it among the components of the Tesla 242 D.

The microscopist returned and, after cleaning all the surfaces with a chamois cloth, began to assemble the microscope. Around four o'clock in the afternoon the microscope stood assembled, but one screw was left over. The microscopist took the microscope apart again and, looking at the diagram even more carefully, assembled it again. At seven o'clock in the evening the microscope stood reassembled and one chrome screw was left over.

Had this happened in our country, the microscopist would have decided that the market relationship of manufacturer and consumer was to blame, that the diagram was for the dogs, and that the microscope should be written off. The American microscopist called home apologetically, took the microscope apart again, and reassembled it a third time. He was not capable of admitting that the diagram might be defective and that a product, bought and tested, might not need all its components.

He worked late into the night and would have continued in the morning, but he came in late and in the meantime the joker had taken the extra screw away. Perhaps he confessed. That I don't know.

In any case, I told Dr. Rossman this story in the darkness where, with the specimen rod held with a match, we searched for lymphocytes in the nonexistent thymus of a thymusless mouse.

After a mere fifty hours of mutual sojourn in the darkness of the former warehouse, at the Tesla 242 D microscope, we found the lymphocytes. In light of that historical happenstance according to which we make discoveries, semidiscoveries, and quarter-discoveries in our laps and not in a luxurious laboratory, as well as in light of the circumstance that everything is one way but also somewhat the opposite, it was all completely logical.

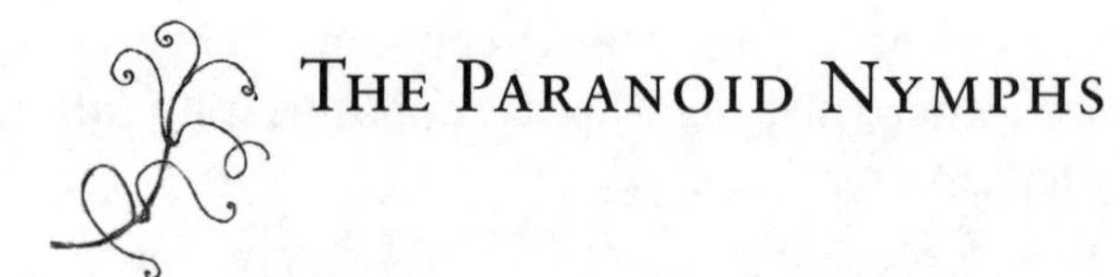

The Paranoid Nymphs

"When the body is immersed in death," says Antoine de Saint-Exupéry, "the essence of the man is revealed." Saint-Exupéry is a man whose essence I deeply respect, both in his life and in his death. He was more than a writer.

But I have some difficulty with the sentence quoted here. It has a strict delimitation: it applies to adults and it applies to people on the Golgotha road of disease or fatal trauma. But what about children dying, not necessarily from disease, just dying before heading down any road, just dying without notice and without knowing what is happening?

There is sudden infant death syndrome. One- to four-month-old infants simply stop breathing in sleep, about half of them with a prolonged hypoxia, oxygen deprivation in the tissues, which may go unnoticed, and half of them without even this warning, according to Finlay and Rudd. The condition has been labeled primary alveolar hypoventilation; the term *primary* usually means that the cause isn't clear. Primary alveolar hypoventilation is characterized by the failure of automatic, involuntary respiration, despite normal lung and chest bellows. The absence of the automatic ventilatory response to hypoxia may theoretically be due to some failure of peripheral chemoreceptor activity, that is, the body not noticing that something has gone wrong. The absence of the carbon dioxide response may be due to hypoxic depression of brain activity or hypoxia-induced increase of cerebral blood flow that cheats the notification systems. In short, the id in the body stops sustaining itself. This is a different id from the one invented by Freud.

And it isn't that the id wouldn't yet be accustomed to life in the crib or elsewhere on the planet. The human fetus in the third month of gestation has already begun breathing movements that last from seconds to hours. At first the movements occur at irregular intervals and last from ten to seventeen hours every day/night period, during active (rapid eye movement) sleep. Before birth, breathing becomes regular. But occasionally, it stops.

Both children and adults can develop a central alveolar hypoventilation due to a lesion in the high cervical cord or brain stem, sometimes due to encephalitis, poliomyelitis, or tumors, or to surgery in the area. Again the involuntary breathing stops in sleep. This syndrome has been given the very poetic name of Ondine's curse. According to a German tale, Ondine, the water nymph, took away all the automatic functions of her unfaithful husband, who thus forgot to breathe in his sleep, and died. In Dvořák's opera *Rusalka* the incident is accompanied by beautiful music; in real life, there is only silence, the deepest silence of all when it happens to a child.

Considering sudden infant death syndrome, the only Ondine I could imagine would be a nymph of the collapse of the wave packet in quantum physics. Or a vengeful nymph, faster than light.

The same nymph has another syndrome in her arsenal: infant progeria, the disease of the minutes on the clock of human life, the disease of fate and of being. At the age of several months or years, growth stops; weight decreases; the skin becomes atrophic, transparent, and wrinkled; the hair thins; the skull becomes relatively large; the face acquires birdlike traits;

the teeth are crowded; subcutaneous fat disappears from the face and chest and extremities; the serum lipoproteins increase, with resulting atherosclerosis; the blood pressure goes up; senile mentality develops; the eye lenses grow opaque; the hearing goes; arthritis occurs; and the kidneys and liver may be affected.

One becomes an eighty-year-old infant. One plays with dolls or with Lego toys, but one hides one's possessions from others. One is suspicious, sour, obnoxious; one does not like to hear any news, even if one still hears. Time goes too fast; more precisely, the internal clock goes too fast compared with the external one. It is called the Hutchinson-Gilford syndrome, sometimes mixed with Cockayne's syndrome, both hereditary and recessive—that is, both parents might have the gene without noticing it. Not many cases have been described so far. But my question is this: does the same paranoid nymph just come a little later to people who have managed to escape other lethal nymphs into old age?

In other words, infant progeria may be an exaggerated form of normal senescence, which is in its own right just a way to tell the individual that enough is enough and the species requires the succession of generations. But infant progeria would prevent even the succession of generations; if it is a disease, it is a paradoxical, wrong-for-wrong's-sake disease.

The human essence, revealed, involves the anthropic principle of the universe as well as the paranoid nymphs of sudden senescence and sudden death. I doubt that the ontologic philosophers know enough about paranoid nymphs and mistakes of evolution. In reality, it is usually very hard to tell who is a mistake of evolution and how many paranoid nymphs are still at large.

From the Amoeba to the Philosopher

The evolution from the amoeba to the philosopher, as the joke goes, is called progress by the philosopher, not by the amoeba. As an immunologist I am somewhere between. Occasionally, I listen to philosophers of science in order to learn whether my concept—what concept? say rather my feeling or sense—that progress exists, based on the change from attenuated and dead vaccines to genetically manipulated "naked DNA" vaccines, is legitimate. After all, only philosophers of science tend to talk about "science." We scientists talk instead about active immunization, about hedgehog genes (regulating the development of mice and men), and about lymphocytes. We come upon the word *science* only when we're on the defensive, under pressure from all those antiscientific or "alternatively scientific" moods, emotions, and visions, where the main issue seems to be not what our findings are but whether we are postmodern enough. Ernest Rutherford, one of the founders of nuclear physics, is reported to have said, "Don't let me catch anyone talking about the Universe in my department."

So we listen to the philosophers of science to learn something about global problems and changes. We may be afraid that our sense or concept of progress could be just the result of our scientific ideology, "just one of many ideologies that propel society," writes Paul Feyerabend, one of the ideologies from which "society and its inhabitants" should be defended. Any ideology, says Feyerabend, that makes us question our inherited beliefs is an aid to enlightenment; a truth that reigns without checks and balances is a tyrant that must be overthrown. Science in the eighteenth and nineteenth centuries was indeed an instrument of liberation, Feyerabend admits, but for him it does not follow that science is bound to remain such an instrument . . .

As a humble immunologist, I shy away from big plastic terms like *truth* and *progress,* worried that I—together with my science—may have become rigid (Feyerabend), and have not even discovered *a* truth, let alone *the* truth. For anyone who has ever lived in a full-blown totalitarian system like Soviet Marxism, ideology is something close to death by cerebral atherosclerosis. Ideology was for us something like a faulty consciousness, something produced not by a brain but by a committee. "I would trust a Shakespeare, but not a committee of Shakespeares," remarked the Polish satirist Kotarbinski.

But here I must inject a note of caution: it is risky business not to differentiate among ideologies. It may have been a certain "ideology," or science, that was the first agent of liberation from official Soviet dogma. Long before the popes of socialist realism and Marxist Leninism could be exposed, ideological "scientists" like Boshyan, Lepeshinskaya, and Lysenko were put on ice and labeled obstacles to the progress of Soviet science, and that by influential and less timid Russian scientists themselves.

But those may have been extraordinary conditions. Nowadays we may still be at risk, with all our immunology and vaccines, polymers and quasars, our "spurious" results of current "rigid scientific ideology." At best, they may be some kind of solution among solutions to the ecological, sociological, and anthropological situation caused by contemporary civilization's disastrous commitment to technology, science, and "inherited beliefs." At worst, they may be a false kind of progress leading only to the status quo ante.

It's like the progress in human well-being and affluence: anything above the threshold of bare subsistence may be relative, subjective, and even dubious if it means progress in which only a few hundred thousand people are left better off instead of billions of them.

Here we are confronted with a basic, I would say cruel, paradox of the present world: scientific technology and Western civilization are not that badly needed in the so-called First World, which produces them. Science isn't much needed by California philosophers and postgraduate mystics. It's needed by those 4 billion people in the Third World if they are not going to perish by famine and kick the bucket in the traditional framework of human values, rights, and aboriginal habits.

But even citizens of the First World don't have uniform viewpoints and perspectives. Scientific progress means different things to a healthy citizen (especially a well-off healthy liberal egghead listening to the prophets of new anthropology) and a patient with AIDS, who would simply regard the naked DNA vaccine as an answer to the ultimate question. Similarly, for a hemophiliac, genetic manipulation may represent absolute progress from zero to one hundred. To a human being who is allowed to go on living and pursuing questions about the meaning of life by means of a heart or kidney transplant and the discovery of cyclosporin A, the problem of interference with the wisdom of nature may appear to be rather an abstract question. From the viewpoint of an individual affected by a genetic disease or a deep metabolic defect, with an irreversible and undeserved loss of key functions, nature doesn't seem all that wise to begin with.

Liberation from scientific ideology is in some situations a very secondary problem; the real problem is liberation from inefficient therapy and progress toward useful causal therapy.

As for Paul Feyerabend, I have admired his temperament, but I don't admire this sentence: "We have become acquainted with methods of medical diagnosis and therapy which are effective (and perhaps more effective than the corresponding parts of Western medicine) and which are yet based on an ideology that is radically different from the ideology of Western science." I regard Paul Feyerabend as one of the most original men on the planet, but I don't believe that he would have chosen something "radically different . . . from Western science" had he contracted a grave disease.

In other words, if we stick with a metaphorically used term like *scientific medicine,* which philosophers know more about than they do about, say, polymer chemistry, the word *progress* will sound different to a robust philosopher in Berkeley than to some poor soul with amyotrophic lateral sclerosis. And a physician must see and accept and advocate the priority of progress (or whatever term you give it) from human suffering to less suffering or an absence of suffering.

Similarly, being an "inhabitant of society," I can't help but call it progress when prisoners of war in many countries are no longer killed, as they were by the Mongols, or mutilated, as they were by the Slavs in the Middle Ages, when slaves are no longer tortured as frequently as was observed even by Darwin, and when victims of tertiary syphilis are no longer burned to death in the marketplace as witches.

I can't help but remember that the concept of progress arose in the seventeenth century, with the refutation of the idea of the decay of the world, with the concept of a human future, with the revelation that the tempo of improvement, innovation, and invention was speeding up, and with the notion that Peter Medawar calls "perpetual plus ultra": that there would always be more beyond, open to human ability. Thomas Hobbes referred to it in *Leviathan* as "joy arising from the imagination of man's own power" and the basic sense that "there can be no contentment but in proceeding." Bacon, in the *Novum Organum*, expresses the same feeling: "I am now therefore speaking of hope. . . . There is no difficulty that might not be overcome."

The idea of progress was originally a science-and-technology-induced idea, and it was emotion-laden, powerful, and organic. William Godwin wrote in 1793, "Can we arrest the progress of the inquiring mind? If we can, it must be by unmitigated despotism."

What has gone wrong with the inquiring mind since Godwin? Did it differentiate into two or more modes, one still fed by the notion and intrinsic evidence of progress, with the corresponding emotional driving force, the other conservative, missing the old, steady "natural world" as a paradise lost, missing the old "human questions" that may be, sometimes, off limits to science? Reading about the "natural world" always reminds me of S. J. Lec's aphorism: It would be marvelous to live in a jungle where the laws of the jungle didn't apply. I have the feeling that some of us mix postromantic, postmodern, ultimately postcommunist reveries into our scientific inquiries. Nobody would dare to argue about the tyranny of education, so we speak about the tyranny of rationalism instead.

There's a train station metaphor at work here: the train, whose operation we don't understand, doesn't arrive on time, so we grow impatient and dream about the old oxcart or the scooter we had as kids; we are impatient and equipped with a backpack as a metaphor. The representative sentence attributed to James Thurber is at least witty: Progress was all right, it just went on too long.

It follows that the mode of progress selects patient and intellectually collaborative human types who understand the continuity of train traffic, even if the train is not in sight right now. They accept the definition of George A. L. Sarton: "scientific activity is the only one which is obviously and undoubtedly cumulative and progressive," whereas "progress has no definite and unquestionable meaning in other fields." These types have as a rule a full sympathy for the humanities, where the experimental method is impracticable but may represent the key disciplines of the future, as was frequently stressed by Lewis Thomas. Provided, of course, that the scholars in these fields stop reacting to the situation with floods of words or with excessive gloominess, not to mention unnecessarily complicated jargons that make comparative studies and the history of single fields difficult indeed.

Only in science can progress be assessed rationally, said Karl Popper, and therefore "only progressive theories are regarded as interesting." Humanists alarmed by the word *rationally* should at least think about biologist Peter Medawar's metaphor: in the management of our affairs we have too often been bad workmen, and like all bad workmen we blame our tools. Regarding the idea that progress is taking too long, one should understand that it is a bit early to expect our grander expectations to be fulfilled. Only during the past 500 years or so human beings began to be, in a biological sense, a success.

We seem to be content with the notion of the legitimacy of our species on Earth (and in heaven). We're not prepared to tackle basic biological questions of survival, dynamics, and adaptation/selection. Little can be learned from nature itself about direction and progress, or about nature itself. I've never been sure about whether our hypothalamus is nature or not, whether our retroviruses are nature or not, and whether selective pressures operating in anthropogenesis and human history are nature or not. Not to speak of the fact that for a rat, nature means a cellar in an apartment building, for a Manhattan cat it means the millions of rats in Lower East Side hideaways, and for a vulture the road with its plenitude of flattened dogs, cats, hedgehogs, and porcupines.

It isn't "nature" that's the agent of direction, but the second law of thermodynamics with its increased entropy and decreased order in closed systems. Life goes in a counterdirection in this case: by way of a series of steps of self-organization, life is order out of chaos and follows the cosmologic rule of increasing complexity in an open system that was originally immensely simple, arriving at our present-day wealth, diversity, and complexity of forms and functions. Only a really progressed form of life can deny that life has progressed in the last 3 billion years, wrote Ralph Estling.

Evolution in general has its roots in the imperfection of mechanisms of conservation, and life learned how to profit from errors, disturbances, and disorders that would annihilate a nonliving, nonreplicating system, Jacques Monod wrote. This may be a step forward, if we could just know more about the direction and counterdirection. So far, in the realm of the living, one can just say that it arose on the lower limit of preservable complexity, and therefore the only open direction is "up," to an increasing complexity, of course with abundant detours and reversals; the principle of diversity in the development of organs, individuals, populations, and species marked by "early experimentation and later standardization" may be a true mark of history, according to Stephen Jay Gould. But, again, it is a bit too early to judge. Life moved from the lowest limit—single cell, which is still the model organism—to ever larger bodies. There was nowhere else to go, and this, Gould wrote in his most impressive book, *Full House* (1996), is not progress but physics. The world as a whole reflects not one inexorable progress but changes in the pattern of diversity and in the "spread of excellence."

Getting larger may at times lead to an impasse, as with the dinosaurs, or with horses, which are now only a "paleolithic remnant" of a group that once had a great diversity of species of dog-sized animals.

But I am not sure whether the evolution of the human brain can be considered just a "physics," and just another phenomenon of the "spread of excellence" that will be matched sooner or later by the evolution of a competing excellence. The cosmic capacity of the human brain may be such a dramatic and decisive evolutionary step for life as was the step from darkness to sunlight and from anoxybions to the protection against and use of oxygen.

In my understanding, the living world has one clear message: that extinction is forever; once we lose a complex experiment in form, it will not arise again. Had not the weird chordate *Pikaia* survived the Burgess decimation 530 million years ago (see Gould's *Wonderful Life*), there would never have been any vertebrates, any mammals, any primates, any people, any philosophers, any consciousness. Just amoebas, sponges, cephalopods, annelid and priapulic worms, trilobites, brachiopods, gastropods—as the names suggest, surrealistic creatures that might have produced more surrealistic creatures but not a dragon, not a Saint George, and not a surrealist. The magnetic tape of life can't be played again. The message is called the irreversibility of evolution, and even an experimenting immunologist is much happier with this term than with *progress.*

However, I suggest applying this principle not only to chordates and dinosaurs and mammals, but also to the development of remedies, to the decrease of suffering dependent on our skills and organization, and to the functions of inquiring minds (which can be stopped only by unmitigated despotism).

I suggest that we are speaking about irreversible evolution, that we are all against unmitigated despotism, and that we still harbor, somewhere in our hearts, the optimism of enlightenment and of joy arising from the imagination of man's own power.

Trouble on Spaceship Earth

Truth

He left, infallible, the door itself
was bruised as he
hit the mark.
We two sat awhile
the figures in the documents
staring at us like
green huge-headed beetles
out of the crevices of evening.
The books stretched
their spines,
the balance weighed just for the fun of it
and the glass beads in the necklace
of the god of sleep whispered
together in the scales.

"Have you ever been right?" one of us asked.
"I haven't."

Then we counted on.
It was late
And outside the smoky town, frosty and purple,
climbed to the stars.

MIROSLAV HOLUB
translated by George Theiner

Trouble on the Spaceship

Buckminster Fuller said that the trouble with Spaceship Earth is that it came without operating instructions. That's why, so often, so many things here get out of control. Usually we worry about human-made disasters and the absence of human mechanisms for control, but the whole story of mishaps aboard Spaceship Earth is not simply a human story. Nature itself does not seem to have been provided with operating instructions. Meteorites seem not to have been instructed, and Earth's crust, with its pulsating volcano nipples and drifting continental plates, has gotten out of hand numerous times in geological memory. Biological species seem not to have been instructed about not proliferating beyond sustainable limits; this means they must be controlled by parasites and epidemics, which amount to disasters from the species' point of view, but can be just as easily regarded as a means for reestablishing dynamic balance on the spaceship.

Five brief periods of mass extinction have been recorded in the geological memory. The Permian Triassic episode involved the extinction of 90 percent of the shallow water marine species 225 million years ago. A recently discovered global warming, 57.33 million years ago, at the end of the Paleocene, uncoupled the deep from the shallow sea ecosystems and caused the extinction of almost half the species living on the ocean floors in just a geological second, a mere 100 years. And the notorious debacle of the dinosaurs in the Cretaceous era, 65 million years ago, affected the most visible part of planetary life. We are not to blame for that, and what's more, we may even be indebted to it, since it triggered the rapid development of mammals. Disasters may not always be detrimental to life on Earth. They may have enhanced the diversification of forms and thus the development of new species. There is an apparent discontinuity of life forms in the paleontological material; there are missing links. At numerous points in the history of life, one may suggest, new species and genera tend to appear in short time spans: punctuation, as Stephen Jay Gould says, to the longer-lasting periods of equilibrium.

This is not to say that disasters, old or new, natural or human-made, are desirable; it is merely to suggest that disaster is—or used to be—a relative term on Spaceship Earth, that we need some sort of punctuation to have continuation, just as we need death to have evolution.

Only in the past fifty years or so have we realized that the instructions for the spaceship are missing and that we, or more exactly our industry, in the broadest sense of that word, may cause something irreversible. And we've begun to realize that this may happen even when human instructions—known as science, though that doesn't sound as user-friendly—have become available: knowledge of the chemistry of Freons, for example, and of the cellular damage caused by ionizing radiation.

Lack of global information is not the only problem that complicates ecological needs and puts them at odds with the practical solutions of industry and the attitudes of the general public. Another problem is that more on the planet is invisible than is visible: only a minor part of our deck on the spaceship, the world of biological forms, structures, and events, is accessible to our senses. Only a few things are truly comprehensible to us: that is not the message of common sense but the message of modern science.

It seems noble to care about trees, buffaloes, eagles, and white rhinoceroses, but the ignoble insects are equally important, forming an even larger biomass than the rest of the animals, while microbes are in a way the decisive component of the biosphere and its food chains. The homeostasis of the oceans would not be disturbed by a tragedy involving the dolphins, but it would by the extinction of plankton. Life-supporting soils depend on unintelligible properties of weathering; on protists, rhizobia, and earthworms that turn over the equivalent of all the soil on the planet to the depth of one inch every ten years; and on complex carbon, nitrogen, and sulfur cycles.

The trouble is that the "greenhouse gases" work far above our heads, our laboratories, and even our computations, meaning that the fear of global warming is supported by theories, but not by the personal experience of consumers, automobile drivers, heating plant superintendents, and policy makers. Not in Europe, not in America, and definitely not in India or China.

The ecology of disasters and ecology in general is less about the obvious than about the extrapolated. But in one way we all experience ecological dangers and disasters. That is in our dealing with our internal landscapes. All multicellular life forms are essentially ecosystems of molecular and cellular entities, societies and structures in a constant process of evolution and homeostasis, tied together by an overwhelming network of signals and background noises. We experience the state of affairs in our internal landscapes daily. And we experience the disasters, impending and in progress, at least once in a lifetime.

There is a direct relation between the external and internal landscapes, and the internal ecology is at risk when we neglect questions of environmental health.

During our student years, European pathologists used to display giant tissue sections from the entire human lung, mounted on windowpane-sized slides. They showed smokers' lungs with and without tumors, but black as a grim urban street out of Dickens. They would compare them to infants' lungs, nonsmokers' lungs, and the lungs of villagers.

It was an impressive display of landscapes, failed ecosystems, and disasters, all within the range of human responsibility and human failure, systems once in the control of owners who may have believed in internal ecology, more or less, but eventually failed themselves.

The learned ecology of human internal landscapes can also be called medicine, though that doesn't sound friendly either. At any rate, the ecology of the external world can learn some lessons from it, as for example:

- Never get emotional when you're facing grave situations; never get hysterical when you're facing disaster.
- You can't restore the old equilibrium, even if you call it a paradise, which is pure romanticism. Try to establish a new one instead.
- Do not intervene if you are unsure; believe in the great power of homeostasis.
- But do not blindly believe in the infallible wisdom of the body or system in question. It makes as many mistakes as you do, more and more as age increases, to the point of self-aggression.
- What mostly gets out of hand is self-control and confidence in the system in question. In the inner landscape, stress, fear, and desperation are as dangerous as any form of superstition.

From the planetary point of view, bodies are just epiphenomena emerging from the continual flow of genomes, generation to generation; and the history of the genome is the only biological eternity. In the long run, the medical history of generations and diseases offers something reassuring to the ecology of external landscapes and external disasters.

Except for the threat of human-made destruction of the entire spaceship, there has never been a disease or disaster that could exterminate the entire human species. Humanity has not only its immune system, memorizing and memorializing all past afflictions, our internal heritage, but also an external heredity, sometimes called culture, sometimes civilization.

In Europe, we are the progeny of survivors of all the medieval poxes, anthraxes, mycobacterial tissue destructions, salmonelloses, and viral or retroviral infections that attacked the populations of our cities the way the Mayan population was attacked. We are the beneficiaries of stronger immune systems; we also developed new medical technologies, along with a sharper sense of the dynamics of history.

So far, there has never been a total, 100 percent disaster. There has always been some statistical probability for escape, some statistical hope.

So far. Consistent ecological concern for all kinds of landscapes and for the preservation of sanity on Spaceship Earth may help to prolong the "so far" a long way ahead. We not only have to be clever, but we must have something important to be clever about (as Peter Medawar said). We have not only to believe *still,* we have to believe *again*—which makes a great difference (as Lichtenberg said).

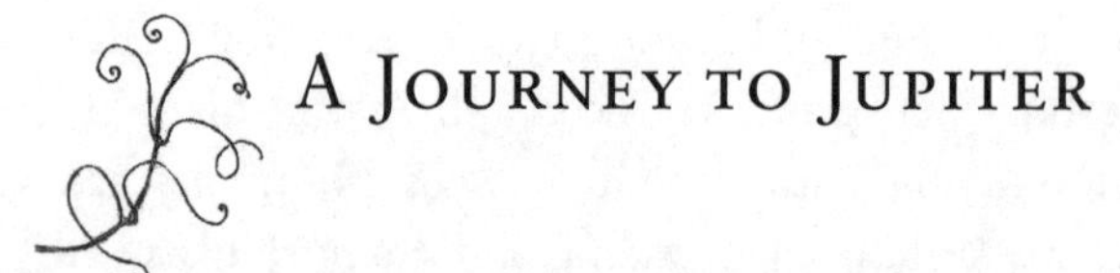

A Journey to Jupiter

A HUMAN BEING'S first true encounter with real nature is not the sun or the lawn or the pet dog. It might be a forest, but even forests, in our latitudes, are too familiar and too tame to be authentic, while in other latitudes they are part of domestic space and the domain of practical realities such as what to kill, what to eat, what to escape from, and what to worship.

The first true encounter between a human being and nature is the night and the starry sky overhead. No technical questions. No practical use. Actually, less use than for ancient civilizations, which had to rely on the sky for tracking seasons, for planting maize, for navigation, and for guessing the will of the gods. Fewer applications, nowadays, and less familiarity. Neither good nor evil. Nothing that can be grasped, understood, felt. Without analogies; just the puzzling void.

I suggest that the curriculum both of the scientist and of the poet begins on a bright night with the megafacts that intrude into the soul's living room. The immense counterpoint of Hoyle's intelligent Black Cloud and the roof of one's parents' house.

I am suggesting that we all begin as astronomers. Most of us simply forget this stage. Some translate that counterpoint into other metaphors, and some forget the megafacts and the century and stay with the magic.

My first report—it was called a "rhetorical exercise" in my high school—was on the canals of Mars. I felt that I was doing something to oppose futility, that I was bringing something into the classroom that was worth being clever about.

I would have loved to see into that black eternity, to see, perhaps, the Martian canals. I couldn't because there was no observatory in my town. So I switched my interest to Darwin and butterflies. But somewhere deep inside I still harbor that adolescent dream that the starry sky is the perfect route to essential reality, to eternity, beyond the limits of ordinary life, death, stone, and candle.

The relation of human beings to the sublime eternity of stars, universes, spacetimes, futures, and even the past is not only the primary one, but also the most tense and intense one. Critical, crisis-ridden.

There seems to be no connection between the sky and us. The essence of the universe eludes our thinking, even when our thinking is armed with instruments and computers. Somehow we miss the target, which flees with the galaxies and their red shift. The Big Bang is a metaphor; even Stephen Hawking is a metaphor.

Since the canals of Mars, I have had just three memorable encounters with eternity. One was on Kitt Peak, Arizona. I was walking around the huge white domed observatory buildings immersed in the blazing Arizona sky, looking like the giant stupas of a brand-new Buddhism reflecting the aboriginal truths that the beginning is the end, that nothing is everything, and that you are also somewhere else and moving at the same time in two different directions. I walked inside and watched these apparently unmanned laser-equipped telescopes and computer-spirited telescopes, like extraterrestrials communicating with the traces of the existing or nonexisting Big Bang, with the cosmic superstrings. I walked outside on the ordinary asphalt roads with traffic signs, through showers of incomprehensible cosmic rays (assuming that I am transparent to the neutrinos coming from nowhere and full of unpredictable atomic events and dependent on fuzzy properties and tracks of particles or waves obeying no traffic signs). I felt—and it was a proud feeling—like an astronomical object, filled with knowledge about lymphocytes, mice, men, and the inner and outer wisdom of bodies. I felt like a tiny Czech cosmic microbe exposed to the cosmic wisdom permeating Earth's atmosphere and hitting the American receptors.

I recalled Alan Lightman's remark about a pious physicist who did some calculations on the origin of the universe that left no room for God and Creation. Asked why he'd left God out of the equation, he replied, "But that's His choice!"

My second encounter was in Oberlin, Ohio, watching channel 25 by myself, late in the evening. For real cosmic encounters, one must be alone. Loved ones make cosmic encounters too earthy, create a sort of tension between silence and Berlioz, or between testosterone and Psyche.

On the screen, line by line, the surface of Jupiter's moon Io appeared, from top to bottom. It was yellowish-gray, smooth, calm, marked by long, shadowy rifts. There seemed to be a rounded protuberance on the rim, surrounded by undulating arms; something like an octopus with an inflated body. Then two gentlemen appeared, watching the same thing and explaining that it was a mile-high volcano crater with lava spewing out to a distance of sixty miles.

Then another picture was set up step by step, and then another. The pictures appeared each in turn, the camera poking around in the undulating wasteland.

It was a live transmission via Pasadena, California, coming from *Voyager 1,* in outer space. *Voyager* approached Io, coming within 18,000 kilometers, and within 278,000 kilometers of the surface of Jupiter's clouds. It was all very quiet but very dramatic. Sitting alone in the terrestrial darkness, one felt depersonalized, just a generic human being, just a minute part of Earth, not necessarily in Ohio but just somewhere on the planet, and *Voyager 1* was some part of the common self, of the common body, sent 640 million kilometers into the deep unknown to touch, with its sensors, the jet stream of frozen ammonia traveling at the speed of 560 kilometers per hour, the immense spherical cloud of electrically charged particles, the belts of intensive radiation, duly transmitting its 800 lines with 800 units each, its 115,200 bytes per second, the entire picture in forty-eight seconds.

Picture followed picture, and one started to identify deeply with *Voyager 1*. Suddenly, *Voyager 1* was one of your limbs, hands, fingers; it might have been cut off, but you still felt the phantom pain. In the darkness, one was closer to *Voyager 1* than to Pasadena.

One felt with Blaise Pascal that being "absorbed by the endless expanse of spaces, about which I know nothing and which do not know about me, filled me with horror." One could imagine Pascal watching *Voyager 1* in 1657: "The eternal silence of these endless spaces terrifies me . . ."

I began to fear that something might be out there, some sort of cosmic monster, Spiderman, Shredder, SuperOndine, or, much worse, just an alien mouth, trap, tooth, arm, foreign hand, or, worse still, just a web, string, hole, magnet, or killing field, something that was going to catch poor *Voyager 1*, with all its eleven devices, cameras, and antennas, and put it into a dark, demonic cosmic pocket.

I realized that the demons we instinctively evoke allow us to humanize our connections to gluons and vector bosons and white dwarfs and quasars and wormholes. Projections of inner space into the outer space. They are not real, but they get us involved.

Nevertheless, nothing happened. No tooth or claw appeared. *Voyager 1* and the gentlemen in Pasadena went on quietly. Now we know that Io is one of at least sixteen moons of Jupiter, has volcanoes or cones with plumes rising 280 kilometers high, has no impact craters from meteorites, resembles Earth, has reddish polar caps and a sheath of yellow sodium clouds. We know that the moons Ganymede, Calypso, and Europa are whitish gray, with a high water content, so that they extrude ice from their cones and rifts. We know that Jupiter, with its yellow, orange, and purple stripes, is buffeted by dozens of permanent storms, including a nameless hurricane in the shape of an oval red spot twice the diameter of Earth, and that it has a ring of dark material 8,000 kilometers wide and up to 29 kilometers thick.

Everything that happened would have been perfectly intelligible to Blaise Pascal. There are no demons here. No cosmic pocket. Just a little draft, the kind you get when you open a door.

However, the universe without fear feels wrong somehow. The universe with Pasadena doesn't feel quite right, and the universe with comprehensible frozen ammonia doesn't feel right either.

Even the universe by daylight doesn't feel right. We need the darkness and Olber's paradox: that in spite of the fact that the combined starlight doubles with each doubling of the space considered, there is still pitch dark after sunset. I am once again enchanted by the solutions of Olber's paradox, as I was enchanted on a bright Australian night, seeing with the naked eye the Large Magellanic Cloud. Watching the black night sky, I witness the age of the universe; we are illuminated not by all the stars, only by the stars up to 10 or 15 billion light years away, assuming the Big Bang actually occurred. The rest of the starlight hasn't arrived yet. If, on the other hand, there was no Bang, we should remember that the life span of stars is somewhere around ten years to the eleventh power, while the limit of visibility is somewhere around ten to the twenty-third power light years; in an eternal universe the majority of stars in the range of visibility would be dead. Extinguished. In 1848 Edgar Allan Poe suggested this in his essay "Eureka," one of the very few cases in which a poet made a real contribution to the understanding of eternity.

In the case of a Big Bang, however, darkness at night is a new phenomenon. There was no primordial darkness. Radiation prevailed, over 300,000 years. The brightness of space—with nobody around to call it a sky—was produced by the temperature of the leftover radiation, which was light enough to produce a primordial brightness. At the end of the radiation era, the temperature was still 3000 K, a pervasive red light.

And that's my third encounter with cosmic eternity. A continuing nightly encounter, powered by the assumption that we do take part in the limited visible universe by our radio programs on ultrashort waves, by our television programs, and by radar signals that are not deflected back from the ionosphere. They are there, so to speak, for the visible eternity. We take part in the cosmos not only by means of *Voyager*s and special Arecibo signals, but also through our everyday communications, ranging from *Melrose Place* and *Seinfeld* to Middle East peace talks and Erich Fromm, from heavy metal to Johann Sebastian Bach. We are part of it, we are out there, and some E.T. with a CETI or Sentinel or Meta project may be wondering by now whether we are that stupid, or that smart.

We are not simply aware of the sublime visible eternity; we are part of it, be it in the Jupiter dimension or in the galactic and metagalactic dimension, all far too big for terrestrial microbes, all beyond reality and anterior to demonology. All pure nature, but already a little contaminated by humans. Perhaps that too is nature's way.

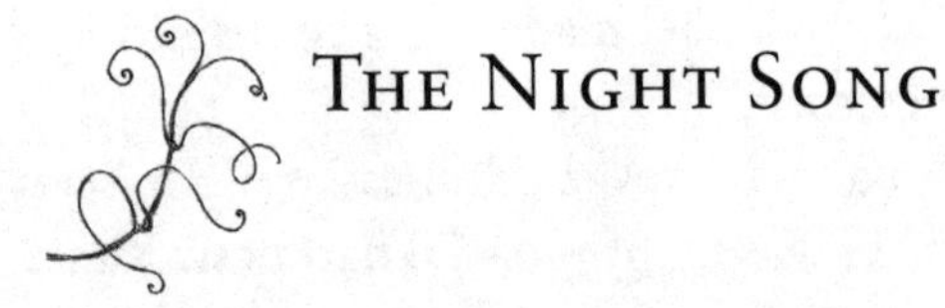

The Night Song

In the countryside I am privileged to inhabit, wintertime is marked mostly by silence. From a distance one can hear the faint rumbling of the city, punctuated by the occasional screech of worn brake linings. Especially after midnight, the silence is transparent and almost solemn, palpably cosmic, and no doubt compatible with the well-earned slumber of the just and the righteous. Since January, though, some kind of song emanates nightly from the treetops or hut-tops of our neighborhood. It starts with a long, wailing, interrogatory howl or cry; shifts to a series of high, squeaky notes; and concludes with an unskilled, unschooled twitter and chirp. It repeats. It disappears and comes again. The sleepers, including the righteous and the just, are awakened. Alarmed, they sit bolt upright in their beds. They go to the window. They don't see a thing. In the morning they discuss the experience and work out a critical response to it.

Views emerge across a broad spectrum. The realists maintain that it has to do with pheasants in distress or, alternatively, with pheasants that are faring much too well. The poetic souls insist that it's at least an owl or a goatsucker (nightjar). Some even propose a bevy of owls or goatsuckers, peering in through the windows. The youngster N.K., whose scholastic achievements are not quite what they should be, always introduces the possibility of a ghost or ghoul, a winter banshee or will-o'-the-wisp.

His belief was strongly supported by a nocturnal accident involving a stinky East German car, a Trabant, that subsequently burned down to a ghostly skeleton. Meanwhile, the skeptical citizenness A.S. holds quite another theory, involving a half-drunk fellow citizen who has a large whistle and who uses it maliciously; such people, she points out, are quite common these days. What's interesting is that this skeptical theory is about as metaphysical as the theories of N.K., since the nearest pub that's open in the evening is about eight kilometers away as the crow, or owl, flies, and it's certainly doubtful that anybody who had made such a trip (there's no public transportation after midnight) would have the energy to whistle, particularly since traveling as the crow flies isn't permitted just now.

There are a few of us who firmly maintain that it is, very simply, a night song. It isn't necessary to research the nature and origin of things in every instance. We may inquire into the origins of things during the day, at work or in our leisure time. The night, however, wasn't made for inquiries. Night phenomena aren't intended for analysis, not least because they aren't sharply visible. The world is loaded with busy causalities; when the day is over, let them sleep. The night is a cause unto itself, a global and all-embracing cause; if we suffer from any shortage of nature, that is, of what is happening spontaneously and going right on without us and apart from us, then night is nature itself and we are engulfed by it, wherever we are. If we counter it by deliberations on causes and effects, we will suffer insomnia and be constantly checking our clocks and watches with the aid of flashlights. And possibly we'll interfere with the night song that starts with the wailing, interrogatory howl and ends with a twitter, so agreeably unskilled in comparison to all that will return in the morning.

It is well known, Proclus Diadochus wrote in A.D. 417, that the man who first devised a coherent theory of the irrational perished in a shipwreck in order that the inexpressible and the unimaginable would remain veiled for eternity. Thus this offender, who carelessly touched the remote face of living things and uncovered it, was reduced to the condition from which he began and will so remain forever, rocking among the waves.

I recall my own shipwreck. I was trying to unveil the essence of poetry. It was in 1959. I was—with my first book published and thanks to an irrational public attack by a high-ranking communist official—a well-known (or notorious) young scientist-poet, and I decided that it was time to receive the wisdom, the torch, the baton, and the night song from my dignified predecessor, Antonin Tryb, professor of dermatology and venereology, poet and novelist, in Brno, Moravia. He was seventy-five then. He answered my letter promptly and invited me to his office at a clinic. I arrived punctually with a deep sense of my mission, passed all the patients and the white coats, and found Professor Tryb's reception area on the second floor. It was empty except for two hibiscus plants in flowerpots, and the secretary. She seemed a little distraught but told me that I could go in right away, the professor was expecting me.

The professor was sitting stiffly behind his desk, head raised somewhat unnaturally. As a greeting, he made a grumbling sound and offered his right hand, supporting it with his left. Then we started to talk about poetry, medicine, and science. The dialogue consisted of my questions and his grumbling, laborious, totally unintelligible replies.

I gradually realized that he had had a stroke the day before, but that he wasn't fully aware of his condition and believed he could be understood. All he had left was the hope that he could still communicate; he wanted to be assured that he could speak. With all my powers of concentration, and all the energy I could muster, I kept trying to figure out what he was trying to express and how I should respond. After a while, I became convinced that we were talking about poetry and science, that I was reassuring him that poetry, dermatology, and venereology were in good hands and would remain so, and that there was no torch, no message, and no night song lyrics to be shared or passed on, that we were both lost in waves of irrationality and were both still trying to maintain that we were not entirely lost and not entirely dead. I had reached a point at which I was making the greatest poetic effort of my life.

We were owls or nightjars, and the messages we exchanged were a wailing, interrogatory howl, a squeaky noise, an unskilled twitter.

Professor Tryb died five months later.

When I hear the night song, in the countryside I am privileged to inhabit (though the landscape is rather ugly), I hear also the message, and the mumbling voice, of Professor Tryb.

What the Nose Knows

As we know, every decent citizen has a soul. Some individuals even have an immortal one. It's all the more interesting, therefore, that nobody knows what it is: it seems to me a kind of auxiliary noun, and a reflexive noun, just as there are auxiliary and reflexive verbs.

In reflecting on my own soul, I've been bothered for many years about (a) neurotransmitters, especially dopamine; (b) endorphins; (c) the limbic system; (d) reticular formation; (e) the hypothalamus; (f) the pituitary gland and its hormones; (g) prostaglandins and interleukins; (h) theories of memory; and (i) the right and left brain hemispheres observed in full activity by positron emission tomography. I fully understand that there's no excuse, in the light of the soul, for any of these terms. But as we know from school, even an excused absence is an absence; if I'm to speak in the presence of the soul, as well as with a soul present in me, I should know about everything that affects it and what the limits of its freedom are or, to put it another way, which of its freedoms are illusions.

I feel I'm a much safer subject for ethology, which examines how I resemble others, than for psychology, which determines how I differ. And in human ethology, Milan Kundera's aphorism applies: even stupidity is the product of highly organized matter.

So in reflecting on my own soul, I haven't really gotten very far. I live in a permanent, hidden, but substantial tension between my ethology and my psychology, between the moves of the soul and the moves of the body, moves that probably shouldn't coexist, but do. I'm especially confirmed in my psychological skepticism by incidents of smell.

I remember one such typical ethological situation from a Czech academic institution. A certain senior technician K. was given the task of weighing out a slight amount of a substance called scatol. It is—like the majority of other chemicals to be found in a laboratory—a whitish powder, in this case methylindol, C_9H_9N, the product of decomposition of the amino acid tryptophan. It has a very distinctive odor, like ptomaines, which are also created during the decomposition of animal proteins and their amino acids, by microbes. In this high-minded definition it sounds completely clean and not the least bit repulsive, but in reality ptomaines and scatol originate not only during the common and familiar decomposition of corpses, but also during the decomposition of proteins from any sort of food in the digestive tract or, less high-mindedly, the intestines and their excrements, and are distinguished by their characteristic smell. I don't know why our olfactory analyzer happens to be so sensitive to scatol in particular, but I can say that it is one of the worst stenches I've ever experienced, and that even includes the Pilsen yeast factory, which used to stink in a way that can only be compared to a blow on the forehead with a nine-pound hammer. Moreover, the smell of scatol is, so to speak, absolutely distinctive and unmistakable.

With a mask across her mouth and holding her breath, K. measured the weight on the analytic scales. When she had reached a tenth of a milligram and needed only a barely visible smidgen more on the tip of the lancet, someone opened the door and the whitish powder, the methylindol, was lightly stirred up.

This was quite sufficient to deposit a nearly indiscernible quantity on K. A stench that was unfortunately even more unbearable than the yeast factory hammer blow spread. It was at the level of simple human excrement but to the third power, that is, as if a motorized infantry regiment had deposited the products of its digested rations on a single pile.

Nobody ever did find out what became of the weighed-out portion. A stink alarm ensued. K. was led out of the olfactory inferno, and a sojourn in the shower was recommended to her. After an hour of showering, the wretched K., known and respected by everyone for her diligence and industry in her declining years, reappeared among her co-workers. She exuded a stench worthy of Rabelais.

How she got home, I don't know. But she probably had to walk, since she was immediately ordered off every means of public transportation she attempted. People passing in the street apparently turned to look at her with an interest she had never before experienced.

I also don't know how things went at home. I am guessing that from that day she must have led quite a solitary life. She did mention later that the neighbors called the building supervisor at one point to complain about the blockage of the sewage system in some part of the building.

Being diligent, she came back to work three days later. Always somewhat shy, she now reminded one of a Psyche who had gone through redistillation and was edited out of the stories and myths of classical antiquity. She stayed in the corner, spoke very quietly, and moved as little as possible.

Nonetheless, after a while the room filled with that unmistakable and unshakable olfactory fluidium that forces everyone to search the floor for its source and doesn't inspire anyone to think about the laboratory decomposition of tryptophan, only about its product.

It was really remarkable, since experts know that scatol, in insignificant amounts, is perceived more as a mildly unpleasant smell than as an incriminating stench. Apparently a not-so-insignificant amount had lodged in K.'s hair and skin. K. was delicately given to understand that full replacement growth of skin and hair takes weeks, and that it would be inappropriate, during that time, for her to take solitary walks in the Pruhonice park after dark. She accepted the advice with the timid eyes of a redistilled Psyche and came back the following week in an olfactorily inconspicuous and unremarkable state, although a bloodhound would probably have found her, even after two months.

I am not a bloodhound and I sincerely respect K., and after the ptomaine incident I felt toward her as I would toward a scorned member of the family, say an aunt: I've always had the feeling that our only chance in life is to stick with people who are discriminated against and ostracized. Otherwise we find ourselves in the same corner as the power-hungry clowns who always succeed because they wait to see who wins and then rush in at the right moment.

Nonetheless, I can't separate the concept of K. from her olfactory existence. Just as I can't separate the concept of my first unfulfilled love from the sweet mingled scents of Chanel perfume and floor wax in the Pilsen community center, nor my first fulfilled love from the moldy smell of central heating in the houses in the poorer neighborhoods of Old Prague. Equally inseparable is my concept of the heart's chambers from the odor of the autopsy room in the Bulovka hospital. Or the concept of ovarian cysts from the carbolic smell of Prague boats on which, for almost a week, I prepared, to put it elegantly, for the final comprehensive exam in gynecology. Equally inseparable for me is the expulsion of the Sudeten Germans from Czechoslovakia from the smell of bread in the bakery of the legless German baker in a Sumava village. Karel Siktanc wrote, "Maybe they're crushing almonds in bakers' heaven," but accompanied by that same scent, they're also crushing the legless baker.

A visit to the stationery store still evokes that feast of paper and glue smells that it did years ago. The metro has a characteristic smell in Prague, another in Paris, and another in London. The smell of forests in Sumava is different from the smell of those in Posazavi. And distinctive gasoline smells will help you distinguish a Berlin street from a Warsaw one, not to mention Mexico City, whose smell is closer to fresh horseradish than to fossil fuels.

There's a characteristic smell to writers' gatherings, of cigarettes, lousy brands as a rule. And an indefinable, delicately pungent trace, maybe of pheromones, in the air of a room where there are many women together. There's a rainbow spectrum of sweat odors from athletes to feminists. The stench of a neglected outhouse used to be the trademark of a Czech or Slovak office and directorate of anything whatsoever. With my eyes open I can never clearly explain how I know I'm in a hallway in Jungmannova Street in Prague, where the British Council used to be, rather than in the Bratislava Virological Institute. With my eyes closed I can tell right away, and could explain it if there were words for those delicate shades and wafts of smells. For reasons with which I could fill two pages, the warm smell of a horse's stall evokes a gala presentation in a Pilsen theater, while anise calls up a motorcycle crash, a half-dead dog lying on the ground, and a cane with a rubber tip on the end.

Why is it that, as we go through the world, we hope to go with our minds open, our eyes open, but above all with an unstuffed nose? What is it that so easily overcomes our intellect and our beliefs and makes us captive to subhuman associations? Why is it impossible to unstitch an olfactory impression, even though we know that it may have been a matter of a minor chemical incident with methylindol that didn't have the slightest connection with the person, the body, or the soul? Why do we know that tryptophan thou art and into scatol thou shalt one day turn, ornithine thou art and into putrescine thou shalt one day, through decarboxylation, return? Why do ptomaines determine our orientation in the world of the living?

It's because of this: although we are not bloodhounds, our olfactory analyzer connects us with the prehistoric life of vertebrates and with those layers of the brain whose existence is incompatible with the public arenas of highly placed citizens like ministers, directors, and chairmen. Those are the same layers of the brain whose existence is grossly incompatible with the soul, mortal or immortal.

Our olfactory system is quite directly connected with the brain's limbic system; this system, over 450 million years old, regulates moods and our readiness to act. It is the network of motivation and emotions, and the processes of learning and remembering are closely and inseparably connected with it. The basically neutral life situation is perceived positively if the limbic system, and particularly the entire right hemisphere—let us call it the pictorial-feeling-musical hemisphere—is governed by euphoria. And this determines the pleasure or displeasure with which we remember a situation. A scent or stench stamps itself on us like the stamp of the official who issued our baptismal certificate.

An olfactory impression carries us directly and involuntarily into the area of inherited, innate reactions. And because from the evolutionary viewpoint, these reactions exist specifically for the propagation of the species or genus, the expressions of emotional reactions are significant signals between individuals of a given species. An olfactory impression and the reaction to it are inseparable from emotions, feelings of pleasure, displeasure, and ill will, and through them they clandestinely but instantaneously control our ideas and opinions.

An olfactory impression is, in the end, one of the keys to

our existence on the planet. What we will miss on travels beyond the limits of everyday experience will be the olfactory aura of this planet, that alternation of scents, smells, and stenches that we struggle through on our way through life. During long-term stays in spaceships, American astronauts missed precisely this variety, because they only had orange-scented napkins. They should have had hundreds of napkins redolent of burning oak logs or the pungent smells of yeast, fern, ferret, and fermentation.

In small animals, olfactory sensations dominate the choice of an evolutionarily advantageous mate: every male mouse gives precedence to a female that differs from him in the system of major histocompatibility complex of antigens over one that is identical, provided he has a choice. This is to prevent inbreeding, which may result in defective offspring, as in some human dynasties.

Our dulled noses hardly perceive such differences in body smells. Thanks to nonphysical cultural heredity, the value range of human scales is altered and alterable, a fact that the perfume and deodorant industry takes advantage of with significant economic success.

Nonetheless, what is left for us of smells and inexplicable or inexcusable emotions is enough such that we can't fool ourselves, on any level of the social ladder, that we're motivated only by reason and morality.

If I wanted to go into private practice in the field of alternative medicine, I would manipulate fragrance and stench associations to the patients' physical sensations. You don't feel well? Here's some dimethyl sulfide. You happen to feel better? Have a whiff of camphor. You feel completely well, sniff geraniol. Geraniol, which evokes roses, can be obtained for a good price domestically, and if it doesn't make you feel better, at least it won't stink up the house. Which is, of course, only a delicate manipulation based on conditioned reflexes. A much more ambitious use of aromatherapy is envisioned by one Mr. Tisserand, who hopes to treat deformed faces and broken fingers with the aid of various terpenes.

"Positive energy" is said to emanate from certain trees that "literally recharge" a person, according to a typical Sunday newspaper supplement article written by a typically supplemental professor. Among the recharging trees, according to the professor, are oak, beech, and all evergreens. It's a fantastic stroke of luck. Oaks don't have to be imported or calculated into health insurance. We simply chase our stressed patients out into the forests. We even save ourselves the effort of diagnosis. You've got aches and pains, go out with your unstuffed nose and get some positive energy.

It's too bad that in our dusky, mythic past, our Czech founding father, Premysl Orac, had a good opinion only about linden trees. Had he known about the German oak, the Premysl dynasty might have enjoyed better health.

According to other Sunday news reports, a perfume has been found that increases playfulness, risk-taking, and childishness; it will be dispensed in Las Vegas. A citizen who cares for his soul shouldn't even show up there to take the air.

For the rest of us, the coexistence of reason and instinct, intellect and olfactory dependence is an inescapable fact. It is also the subject of a permanent inner dialogue among two or more selves. We are pulled along by hidden emotions, fragrances, and smells. We're held captive by ptomaines, but we realize that it is so. We sport a highly characteristic nose on our face, but we don't have to follow it blindly.

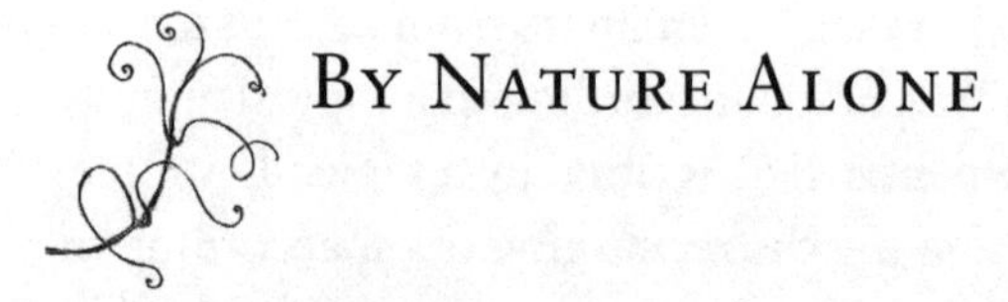

By Nature Alone

As the social contract gets more and more complicated, and as we grow more dependent on microprocessors, regulations, the functions of the higher nervous system, and aspirin, Jean-Jacques Rousseau wakes up in our souls, through some feedback effect, and wails, "Everything is good as it came from the hands of the Creator of the world; everything is ruined by human hands."

Naturalness and natural development, untouched by civilization and science, are good in themselves, says Jean-Jacques. Unnaturalness is the root of all evil. Man would be good by nature, but social organization, based on the protection of property, along with the "enlightenment" it requires, destroys him. There are too many experts, researchers, and artists, and too few simple human beings. We need more nature.

Although he was a founder of pedagogical science, Jean-Jacques sent his own five children to an orphanage. He had a weak constitution and the children were probably annoying.

Weaker constitutions have a tendency to appeal to elementary humanity, mainly because nobody knows what it is. It's probably the above-mentioned naturalness, a sort of naturalness that lies outside time. Rousseau himself said, quite aptly, that among all the human capabilities, reason develops last and with the greatest difficulty, probably because it is no more, so to speak, than a composition or alloy of all the other human capabilities. So in many careers we never even work our way to reason and prefer to remain in that elementary humanity. Logically, one could also call it nonreason or prereason, but that doesn't sound positive enough. "Elementary humanity" is more balanced. Perfect reason, said Molière, avoids all extremes.

What is natural and what is less so, or completely unnatural, is a matter of convention, and a very unstable convention at that. In Rousseau's day it was as natural to have pigpens and sewers in the street as it was to have aromatic herbs, sheep, and shepherds in the countryside; it was natural to wander off and see the world, to meet werewolves, to use candles for light, to poison an enemy with arsenic, not to mill flour too finely, and to treat syphilis with mercury. These natural parameters of life naturally had a rather significant influence on its quality, and thus on the basis of opinions about life, including what is natural. Nonetheless, it wasn't very clear, even to Rousseau, whether the hands of the world's Creator produced mercury, arsenic, floods, candles, boots, shepherds, classicists, and werewolves, or merely the chlorophyll in leaves, and the higher vertebrates.

Nature itself, then, though unchanging in forests, hillocks, meadows, and surf, was rather changeable where the inner landscape of the body, inhabited by microbes, signaling systems, and immune reactions, was concerned. So perhaps there was only the soul left to be natural. But of course the state of the world is surely also part of the contents of the soul, and the naturalness of the soul is a function of the naturalness of the world, and so on, round and round, endlessly, so that Molière's perfect reason had to keep its distance even from the extreme opinion that something can be natural from the cradle to the altar, from the Apennines to the Andes, and from Anaximenes to Newton.

Nature's intentions are not identical with human hopes. Nature cares nothing about individuals but acts only in the interest of the species, that is, the collective genome.

The immune defenses of any mammal are based on an inherent "knowledge" of millions of antigenic specificities they may encounter, and also on the collaboration of a special system (MHC or major histocompatibility complex). MHC helps bind peptides from any foreign material that enters the body. Its effectiveness in mounting an adequate immune defense varies in individuals, according to the nature of the invader.

Therefore, during major epidemics in mammalian species, some individuals will survive the given infection, others not. The species must be prepared for surprises produced by the high mutation rate of viruses, bacteria, and parasites. The enormous polymorphism of MHC gives a species a fighting chance against as yet unknown dangers.

As individuals, we are healthy only in a relative way, depending on how well we resist the next pathogen that comes along. "Natural" health is random, unequal, and unfair. It has nothing to do with warmly humanistic pictures of health.

Nonetheless, we still more or less believe, depending on whether the dominant cultural mood is poetistic or scientistic, that anything from nature is healthy, while anything synthetic is less healthy, or even toxic, alienating, and treacherous. Hay fever is natural, and antihistamines, the cure, are a bit of devil's work. We should consider that histamine can be found—despite the unnatural name—in our very cells, and therefore is undoubtedly natural.

A poem is the natural use of signs, while 2,6,8-trioxypurin is the icy breath of an alienated syntheticity, although it's the name of uric acid, the product of our metabolism.

And as long as we're in this area, urea itself forms an interesting divide between the natural and the synthetic, even the organic and inorganic, for in 1828 Friedrich Wöhler synthesized this billion-year-old natural substance from completely inorganic components. That inaugurated the era of organic chemistry and demonstrated that an insignificant change in one molecule, even a mirror image of that molecule, determines whether or not that particular substance is benign or poisonous, sweet or tasteless, neutral or addictive, natural or synthetic. Something as terribly synthetic as nylon is, on the molecular level, a polymer tremendously similar to the silk from a mulberry worm's workshop: amide and carbonyl groups, hydrogen bridges, structures of folded layers.

As a rule, a natural active substance differs from a synthetic one in that it contains more inactive or unstable admixtures. That doesn't necessarily play a role in the case of vitamin C, but in the case of a hormone treatment it makes the difference between telephone connections and smoke signals.

Good and natural antibodies—that is, protein molecules bearing specific variable regions that recognize and bind antigens, antibodies from good little animals like rabbits, horses, or goats—have, if they are introduced into the body other than through the mouth, a potentially negative side effect in that, through repeated applications, they may cause serum sickness. However, if these proteins are artificially changed by joining part of a human molecule to the active part of their molecules, their negative side effects are suppressed. We're speaking of the humanization of an antibody. There is no doubt that the humanization of an antibody is a deeply artificial manipulation of a natural molecule; nonetheless, the word *humanization* is used here with the full weight of its meaning.

Roald Hoffmann, who is a Nobel laureate in chemistry and a good poet (who therefore can't be *too* alienated, chemically, from elementary humanity), lists seven factors that make us inveterate, even unnatural partisans of naturalness:

First, a kind of romanticism, a yearning for something that no longer exists or never did. Tending sheep evokes lute music and dancing fairies only for someone who's never herded sheep. A horse stall smells sweet only to someone who goes there to admire the shapely rear ends of horses, not to the one who mucks out what drops from them. The pastoral is an aesthetic category for aristocrats and city folk, not for stable hands. An old-fashioned train is an attraction for those who have time to kill; those whose lives are at stake prefer helicopters.

Second, there is social standing. If new materials are too common, too cheap, too available to everybody, then they stop being elegant: plastic panels on the walls, Dacron in shirts, Promise margarine on bread. The better sort of people will prefer "natural" brick, cotton, ironing, and real butter. The better sort of people in this sense are not only those who can afford it, but also those with a natural gleam in their eye that comes from knowing that they know more about everything, wear naturally worn-out sweaters, and use only organic pesticides in their gardens.

Third, Hoffmann lists alienation. Standardized mass production feels alien to us because mass production is based on rational planning. But we miss the mark of a human hand, the evidence of human error, and the imperfection of human understanding.

That's why someone with a bronchial infection would rather burn his back with a horseradish compress (which is simple, natural, and human when our mother or grandmother applies it) than control the infection with mass-produced capsules of ampicillin, which no longer bear a human trace.

Fourth, there is pretense. Since manufacturers are well aware of the factors of alienation, romanticism, and nature lovers, they manufacture everything possible to look like wood, like brick, like carrots, like black wool on a white ram. Aspirin must taste like lemon; corn must taste like bacon. The thirst for the natural is of course largely motivated and justified by this catering to popular fashion and current tastes. This kind of imitation doesn't aim for substance but only for appearance. Basically, it is greedy deception.

Fifth comes scale. Since ancient times we've been used to considering God's gifts as valuable or scarce; we praise and appreciate what is not given in abundance. When there are too many polyvinyl chairs, too many imitation leather backpacks and polystyrene cushions and color reproductions of famous paintings, we don't feel good. Mass production in communist countries succeeded in burying the historical Lenin and Stalin under an avalanche of cheap busts and figurines.

Sixth, the spirit. That is our relation with the inner self, which we feel strongly about and often call the soul. We like what is in harmony with our nonstandardized, tireless, brave, erring, and unpredictable soul. The essence of the synthetic, scientifically worked out and technologically ensured, is the balance of cause (or action) and effect (or result). The essence of the natural is the capacity to act unpredictably, to struggle through circumstances, to break rocks and overcome the limits of the possible. The growth of algae in cultivation tanks is soulless, even though someday, somewhere, we may have to resort to it as a source of protein. Saxifrage and mountain pine have brave souls, even though they'll never be of any other use to us, provided some crafty fellow doesn't come up with the idea that an extract of saxifrage can be used for gallstones.

And finally, there is fear. Science and technology have rid us of our fear of lightning, of the tooth and claw of nature. But because nature has given us the gift of fear, we have to do something with it. So we use it, with great and green diligence, against sciences, technologies, and our own selves. For a shepherd in the Carpathian pastures, a howl from the depths of the woods is a source of fear. A shepherd in a bus is afraid of the rattling of the motor, the stench, and other phenomena of civilization. It's not an irrational fear, but it's mainly a fear based on sensory impression, not on the basis of knowledge. If the fear were based on knowledge, a shepherd in a bus would have to think about viruses, stress factors, and the smoke produced by the bus and the people in it.

Noblest and greenest is ecological fear about the fate of the environment. The external landscape is discussed in a room full of cigarette smoke, meaning that the inner landscape, like the moonlike countryside in North Bohemia, is taking a beating. Because we can't see into the landscape of our blood vessels and lungs, we don't turn against civilization in ourselves and in our self, for this is a personal self, concrete and not collectively blurred by group sensibilities. In relation to the inner landscape we are unusually brave, whereas the external landscape gives us goosebumps, as if it were beginning to collapse on us right now. Of course it has been collapsing on us since the Middle Ages.

But we've gotten used to sixteenth-century dams and artificial lakes, and therefore we don't fear them anymore. We consider them natural. We've gotten used to the devastation from products of solid fuels. We consider it natural. Hardly anyone visits a museum of pathology and looks at a sample of a city dweller's lungs from the beginning of the century, when the automobile was still something of a unicorn and ladies didn't smoke. Nuclear power plants are rare and unknown; therefore we have to fear them as if they were lightning in a can. Radioactivity created by people is a horror. Radioactivity in a piece of granite is fine, because it's natural.

I once had the honor of visiting the uranium mines at Příbram. In the warm underground corridors miners sat, here and there, on piles of broken-up active ore, eating their substantial lunches. "Is that inactive?" I asked, pointing at the pile. "I don't know," the miner always answered. "My Geiger counter isn't working." Upstairs the mined ore was evaluated for radioactivity—that is, financial value—by an elderly woman who sat, virtually without protection, right by the conveyor belt. We asked the guide about her health. "Oh her," said the guide, "she's already sterile anyway."

Everything in the mine was completely natural, especially the uranium ore. Only the Geiger counter was unnatural. Thus, of course, the high number of malignancies was natural beyond the reach of ecological activists. For one thing, it was all going on in the dark, and for another, there were higher interests involved. And mainly, a miner tends not to believe in radioactivity as a problem. It comes from nature and it's invisible.

Since that visit, naturalness hasn't been particularly dear to me. Especially not ahistorical naturalness. In the same way, artificiality and syntheticity for the sake of state or company profit aren't dear to me; they have to be connected to a purpose that involves a higher quality of life and a better level of understanding.

William Carlos Williams, physician and poet, wrote: "We have to get back to a certain scale, a scale corresponding with our time and not with a logic so rotten that it stinks."

Symbiotic Tranquility

Green plants are the component of nature that symbolizes its praiseworthy qualities, its familiar tenderness and fragrance. Without them it would not be the good old outdoors; it would be more like a concrete container filled with hyenas. I think we cling to this bucolic view of plant life because we can't identify with green plants and hence don't know what their everyday life is like. We don't understand these humble producers of oxygen, these quiet and diligent little pumps of the planetary thermostat.

They put carbon dioxide into the soil, eroding crystalline rocks with chemical skill; they manufacture hydrocarbons, sugars, gels of silicic acid, nitrates, phosphates, calcium ions. They rid the atmosphere of carbon dioxide, staving off the greenhouse effect. From the planetary point of view, they are the only real good guys, heroes and patient servants of the living Gaia, beings without claws, teeth, or blood. They sit calmly, silently, filled with green optimism, enjoying their self-sacrifice.

But when I really look at a meadow, I am not so sure. I'm not sure that it isn't filled with battle screams, piercing cries of hate, terror, and pain, individuals and tribes fighting for nutrition, for light, for space, for carbon dioxide, for bacteria, for fungi; that it doesn't echo with the howls of the winners and losers, the songs of the nascent and the hymns of the dying—

nonaudible, vegetable cries. I'm not sure that the tender velvety mesh of branches, roots, bulbs, and stems is not really an interminable wrestling hold; that there isn't a perpetual chemical warfare among roots, among rootstocks, and among seeds; that there isn't some limitless aggression of the strong against the weak, sick, humiliated ones—all of which is obscured in our delusion of a great symbiotic tranquility, behind the veil of a harmonic biocenosis.

And all this takes into account only the good green autotrophic plants, those that can get all the energy they need from inorganic bread alone, can produce sugars from carbon dioxide. But there are half parasites and full parasites among plants, not to mention algae and other microorganisms. Plant parasites have special organs (haustoria) that they attach to their host's body—and suck and suck. Of course, not in symbiosis but in tranquility, in our view. Not only in green, but also in appalling pallor: like the common broomrape *(Orobanche):* no leaves, just an obscene stem and flowers the color of decay, desolation, despair, detritus—or as a Czech botanist put it, flowers like the underwear of an ancient prostitute.

And there are tender plants, here and elsewhere, that take their tender green from completely enslaved fellow plants: the quiet, philosophical lichens, fungi that have trapped tiny green algae, robbed them permanently of their assimilates: the deepest biological humiliation one can imagine—and at the same time, contrary to our view, a beneficial way of life for both.

The main reason for our view is of course the silence: we may notice the fragrances, we may even notice the communication of plants by simple hydrocarbons, such as acetylene, but they make no noise, except for the soft whispering of leaves. Imagine plant voices, not only the battle cries but even the groans of the dying lilac or the lily of the valley in the vase: imagine the moans of mating aspens. We appreciate and praise the nature of plants because we are on another wavelength, in another realm, on another stage; we don't perceive their kind of drama.

I don't deny that this more dramatic view of the tender life of plants occurs to me most often when I'm working in my garden. The garden, of course, is as a rule very far from a harmonic biocenosis, and it shows it on every occasion. Try, for instance, to eradicate couch grass from a flower bed. This action, the liberation of a rose from the clutches of that anaconda of a grass, equals the struggle of a mortal with an angel (and in this fight, the angel is blessed with an ability to drop tiny embryonic angels from each part of its fractured body). I am unable to regard the couch grass as a humble producer of oxygen or a green optimistic pump improving the household temperature of the planet.

Or try getting your asphodels into one bunch in a corner of your lot. I suspect Edgar Allan Poe never tried, which must be why he had so many nice things to say about this member of the lily family. Asphodels and narcissi will always resist your plans by hiding their aboveground traces, preferably the night before the planned transfer. So you never get them all and your gardening, as years go by, resembles a hunt more than a bucolic Virgilian idyll. Looking at my asphodels and daffodils I usually hear the beautiful lines of Robert Lowell: "While we listen to the bells— / anywhere, but somewhere else!"

As a struggling gardener I detest the combative and vicious activity of plants, but as a negligible human individuum I have great admiration for the brave behavior of couch grass and timothy and thistles, which carry on with a victorious song through abandoned, half-neglected, and senescent gardens, stomping over the decaying bodies of the feeble, intellectually pleasing cultivars. I have a high esteem for the athletic fitness of dandelions, nettles, wallcress, and bindweed, and I recognize certain traits of sorrels and plantains as virtues, even though I have to fight them in my lawn, which may be on the whole my shortsighted version of symbiotic tranquility.

I was recently pleased to learn that optimal conditions in the symbiotic tranquility of plants can produce, in terms of one single grass's growth and ecology, an orderly chaos, one that follows all the mathematical rules. Two Minnesota botanists, Tilman and Medin, have shown that the American wild grass *Agrostis scabra,* called pant-creeper for its sticky, thorny seeds, will grow steadily from year to year only on less fertile soils. In richer soil it undergoes oscillations from year to year and in the soil richest in nitrogen it displays a chaotic behavior, growing abundantly the first year but in three years crashing almost to zero, as a result of litter from the fertility of previous years, which prevents new growth.

Thus, even if there is no battle within different species, there are ecological rules that lead to disaster rather than to tranquility: green optimism is intimately bound up with self-sacrifice, and my lawn resembles my life in that both my lawn and I are having to manage on less fertile soil in order to prevent chaos.

In any case, nothing like the unconflicted green vision of the parks service, the forest service, and the garden clubs really corresponds to reality. There's no reason to endow plants with a tenderness that is lacking elsewhere.

At best we are looking at fights, sacrifices, chaos, and pant-creepers. On the other hand, at our own human best and from the global point of view, we may be part of a great planetary symbiotic tranquility, at least when we're observed from a great distance.

Amidst our own chaos, our cries and screams, our wrestling holds, our battles, alas, we don't see that.

Beasts and Freaks

Mythical animals (beasts and freaks, that is) are cataloged in various bestiaries and dispersed throughout orally narrated or written fiction. They originate for various reasons, especially to strike people with awe, but also—to put it bluntly—because it's a kick. Oddly enough, their origin is extensively dependent on so-called observation of natural phenomena, mirroring on the one hand the amount of imagination involved and on the other the amount of stupidity of the epoch in question. Beasts and freaks rarely reveal new or unprecedented features. They are made up of existing but unrelated parts and appendages.

Beasts and freaks occur either as unique individuals or as categories with few specimens: the Minotaur; the Hydra, with nine self-duplicating dragon heads and a snake's body, a very environmentally unfriendly creature; Cerberus the dog; Cronus, the childivorous deity; Titans, in clans or ethnically pure populations; sirens; lotus-eaters; Cyclopes; multiheaded demons; Gorgons; Laestrygonians, the man-eating ogres; satyrs; Harpies; and Pygmies. It is inadvisable and politically incorrect to confuse these Pygmies with the African Pygmies, or Chans.

In terms of cultural history, there are many sorts of

monsters, such as the Sphinx, two phoenixes, centaurs, griffins, and the manticore with three rows of teeth but a flute-like voice, luring victims. There is a Ganesha and Ch'i-lin, a unicorn with a deer's body and gleaming scales. There's the highly hybrid Chimera and a variety of angels, followed by the devils derived from them, including hellhounds, fiends incarnate, Lucifer, and Mephistopheles. I used to have several subspecies of devil in my puppet theater. I distinguished inhabitants of hell by black faces and red clothing, as opposed to earthly nasties, who had yellow faces and green eyes. Mephistopheles looked prominently Indo-European, probably because of his relationship with Faustus.

There are lots of other creatures with local and folkloric significance. We Czechs have the Cunning Little Vixen (who is usually interpreted these days by a soprano of corpulent dimensions). Czech landscapes abound with an enviable quantity of water imps, fairies crossed with does, amphibious nymphs, sylphs, frogs containing enchanted aristocrats, will-o'-the-wisps, witches, sorcerers and sorceresses, usually equipped with talons, and a multitude of dwarfs and elves (only the Scots and Norwegians can compete in this category), some monstrous fire-dogs, cave-dwelling dragons, and prairie bugaboos, usually lacking a trunk but with great big noggins. We can also boast of naturalized vampires and Czech-made werewolves, arising not only from the fantasies and delusions of untreated folk psychopaths, but also from the sight of porphyria patients running loose.

Classic and folkloric beastly freaks display a rather narrow range of variability. They seem to be susceptible to goggle-eyes, sometimes burning and destructive. They tend to have hairdos styled from snakes, to suffer from excessive or stunted growth, to display protruding dentition, and to possess volcanic interiors, as evidenced by flames coming from the mouth.

They are often composed of attractive parts—a nice-looking girl and a useful fishtail, for example—that, put together, provoke consternation and trouble.

More variety turns up in the category of intellectual bestial creatures; play is their raison d'être and purpose. James Thurber, in an essay called "Less Alarming Creatures" (in the book *The Beast within Me and Other Animals)* depicted and named a host of new animals for a modern-day bestiary, devised according to Lévi-Strauss's "universal mythmaking logic." I once translated this piece for a revue called *The Universe,* as an antidote to the then dominant and grim communist mythology that tended to fill its pages. I wouldn't do it today; these days we have the mythology of alternative science, which disseminates its fun not just for amusement but to confuse young naturalists as well. Still, I should cite a few of Thurber's zoological innovations: the Femur, with a small head and huge eyes; the Metatarsal, with a small trunk and six pointed protuberances in its back; the White-faced Rage and the Blind Rage; the Aspic, a hymenopterous insect sitting on the stem of a Visiting Fireman; the Bodkin, a sort of mouse with the ears of an African fox; and a couple of martinets, looking like emaciated flamingos, dancing with their heads turned up.

It is obvious that Thurber was inspired by the facts of language and the facts of organic nature, which gives rise to my last and best monstrous category, one that can be culturally historic only in a very special sense.

In his essay "Some Biomythology," Lewis Thomas speculates that a Peruvian deity, overgrown with vegetables, snakes, and wings, and in charge of home economics, is in fact a toned-down imaginative version of a two-and-a-half-inch weevil from the mountains of New Guinea that has dozens of plants growing on its carapace and rooted in its chitinous body, plus a whole circus of mites, rotifers, nematodes, and bacteria. It looks unappetizing to predators, and so it lives a long untroubled life, extended in myth. The protozoan *Myxotricha paradoxa* is composed of itself, along with cilia consisting of microbes, spirochetes anchored in organelles (which are in fact other microbes); the assemblage makes an animal, although in principle it is a Mafia and a monster that has not managed to solidify its own identity.

Another protozoan, *Blepharisma,* is more coherent in terms of exterior, but its *inside,* as a philosopher might say, is a chimera par excellence; it possesses three different sets of self-duplicating nuclei, with the DNA in each set serving a different purpose. A macronucleus is for regeneration, some micronuclei are for reproduction, and the origin of cilia is coded in other, tiny nuclei. One set of genes in the *Blepharisma* produces a pigment that swiftly kills the animal if it swims into sunlight. The membrane, coded by these genes, disintegrates and exposes its carrier to the tragic fate of an albino. At times of famine a single *Blepharisma* will begin devouring its fellows and enlarge to gigantic (mythical) dimensions. *Blepharisma* is a myth by structure and destiny; by comparison, any legend of antiquity or modern times is trivial.

But let us proceed in biomythology. Each silly cell of our body, when grown in culture, is ready to merge with, for instance, the cell of a mouse. This produces a monster with two genomes and a fate that might upset Lévi-Strauss, but is beneficial for current immunological weaponry: the hybrid yields monoclonal antibodies. And the entire human body, including those of postmodernists, draws energy from oxidative phosphorylations that take place in cellular organelles called mitochondria. These used to be microbes that had settled in the cells, giving up their own evolution, and now they are plugging away on our behalf, despite preserving their own DNA with a circular molecule. Even we, therefore, have our own interior myth, a bit monstrous, inherited from our mothers (only 0.1

percent of the mitochondrial DNA can come from Dad). The fact that we are squeezed in the merciless grip of aging and deterioration could well be caused by defects in the mitochondrial DNA, provoked by free oxygen radicals, which originate from the protection of our inner Troy from the attacks of microbial and other perverse Achaeans. This is a drama worthy of Homer. It is a fantastic possibility, in other words, fantastic fiction. All we have to do is admit it and understand it, as opposed to fooling around with dragons and black princesses (as captured by—thanks to her particular mitochondria—our beloved woman writer Bozena Nemcová).

If we want something really fantastic, curious, and singularly bestial, we can peel at the carbonized remains of creatures stored for 550 million years in the Burgess Shale, in Alberta, or at the ferrous sediments in the Mazon Creek formation in Illinois, some 300 million years old. *Hallucigenia,* as the name itself tells us, is a surrealist's nightmare, something like a pincushion wobbling on seven pairs of thorns, with a bucketlike head and protuberances on its back reminiscent of the prehistoric stegosaur. This tiny monster—only one and a half inches long—is terminated by a long anal tube, erect as the exhaust pipe of a heavyweight truck. *Tullimonstrum* from Illinois is an appropriately named phantom from a six-inch-long banana gone bananas. Yes, only here can we see genuine imagination, as opposed to a mere regrouping of given forms.

In view of the fantastic features of so-called nature, long past and buried under global disasters, invisible to the naked eye, in view of the fantastic features of the giant (Polyphemus, no doubt) of evolution, I do not find myself bowled over by Tolkien. He reminds me more of the imagination of scribes under Emperor Rudolph II. And in view of the classic legends of antiquity and the more recent intellectual games and lyrical fun, starting with Swift, continued by Lewis Carroll, and recently championed by Thurber, I regard the self-duplication of present-day video monsters as the feeble products of impoverished imaginations, which cannot even design an ordinary dwarf, not to mention a dodo or a couple of enchanted mitochondria.

The Death of Butterflies

Among animals, butterflies are the most beautifully dead. Even their hairy little bodies will survive for a relative eternity, only a little shrunken, while the wings are spared with their original play of color and iridescence, especially if the funeral takes place on boards of linden wood, where little pins fix them in the most entrancing positions. There is nothing finer than a butterfly collection, so long as, within us, the natural scientist is fused with the lover of angel wings.

For this reason we begin a butterfly collection one stage before we "praise the Lord in the company of a girl," as the Czech poet Frána Šrámek would say, the first girl, the second, the fifth. Anyway, who would run around with a butterfly net in a sunny glade when a girl lies resting in the grass, hands behind her head as she reflects on ovulation cycles?

At the relevant stage of my life I gathered a large collection of butterflies and moths, with particular emphasis on tiger, silkworm, and regal moths, marking down their precise family, genus, and species according to Joukl's key and a record of the exact time and place of capture. The collection was the focus of my life, two steps east of Eden, filled with the fiery breath of William Blake and the cool murmur of the waves beneath the keel of HMS *Beagle,* in which Mr. Darwin sailed.

Of the collection two steps from Eden there remained, after

twenty years, only the little pins and the labels with the names of the species on them. The beetles of time ate them up, those angelic creatures. From what had once been Blake, only Darwin remained. I am tempted to say that Blake grew up and turned into Darwin, but that might make romantic neo-Blakes angry.

Jaroslav Seifert would not have been angry. Jaroslav Seifert wrote that whatever we have loved we will continue to love, and I continue to harbor a fondness for butterflies.

I can only admire the bilateral symmetry of human fate that led me to England as a guest of Trinity College, Cambridge, where I attended an evening banquet with students and professors, presided over by V. B. Wigglesworth (who died just recently), the physiologist of insects, whose great work was my bible during that butterfly phase. Before dinner we all rose to say grace. I knew only a few short prayers my mother had taught me, so I held instead a private requiem for a caterpillar into which I had repeatedly tried, following Wigglesworth, to transplant corpora allata, but they always died. I whispered to myself a verse by William Blake that my reason refutes but some deep belief holds on to: "Kill not the moth nor butterfly / for the last judgement draweth nigh."

I can also admire the bilateral symmetry of the fate that recently led me to take part in an excursion led by Professor L. P. Brower to Sierra Transvolcanica in central western Mexico. From November to March six different hills are the home of the first to sixth swarm of monarch butterflies *(Danaus plexippus),* which are attracted over a distance of 1,800 to 2,500 miles from Canada and the east coast of the United States. There are roughly 25 million monarchs on one peak, packed together in the branches and on the trunks of the Oyamel firs, *Abies religiosa,* that is to say, on some 2,000 to 2,200 firs on one hillside. Huddling together, close to the fir needles, the branches, and the bark of the trees, protects the monarchs from humidity and temperatures below four degrees Celsius. From a distance, the firs, colonized by the monarchs, blaze a brown-orange color. If we saw it in our country people would say the trees were dying; here, the orange tree among its green fellows signifies a double life: one tree and 12,000 butterflies.

Just why this fragile butterfly, ten centimeters of black-veined orange wingspan with a tiny black and white chess-board pattern on its tip and edge, should drag itself from Canada to Mexico is such a pleasant enigma that it fascinates both biologists and creative writers.

Monarch butterflies migrate (as Lincoln Brower found) probably because winters became dryer in their native habitat and the plant on which the monarch caterpillar feeds, the milk-weed *(Asclepias* gen.), began to move north. The Mexican monarchs fly off in March, north toward the milkweed (of which they know nothing, however). Diligent procreation begins in Texas and Florida. By diligent I mean that one, two, or several males inject a spermatophore, a little capsule of sperm, into the female. The spermatophore contains a nutritious substance that the female values to such an extent that, after the arrival of a sizable spermatophore, she mates again only after a lengthier period, giving the stronger males a genetic advantage, by the beard of Darwin. Guardedly, the female lays her eggs, one egg to one milkweed plant so that each caterpillar can sustain itself and grow up quickly, and transform itself into a chrysalis.

The monarchs that emerge migrate further north in the United States and on into Canada, where they continue to multiply, so that three to four generations are produced before August comes and they begin to reflect upon their return. The reflection consists of the hormonal suppression of sexual activity and the accumulation of energy reserves in the form of fat bodies in their abdomens. Sugars converted to fat are acquired from senecio, which flowers continuously in the south, and from aster and goldenrod plants.

They give the butterflies strength to fly, day after day, on and on, though no one knows exactly why, since what is it exactly, home? And what is it exactly, returning? And what is it exactly, destination? This is simply how they have evolved, and the monarchs can afford this lifestyle because of an advantage that other butterflies no doubt envy: birds don't gobble them up because they contain glycosides from the milkweed (similar to digitalis), which are poisonous to birds in given concentrations. This is an advantage that another *Danaus* butterfly, the viceroy, makes use of; without having partaken of the glycosides, it evades consumption because, thanks to Darwinian selection, it looks exactly like the monarch.

All the monarchs fly south in September, and by December all of them have left Canada and the eastern United States and have arrived in Mexico, where these great-great-great-grandchildren of the butterflies that made the long journey in the spring have never been before.

So here they are. Twenty-five million fragile orange angels with their conspicuous secret. It's certainly a feat that takes your breath away. The location is at an altitude of 3,500 meters, and you make your way here on the top of a run-down truck, on a road that resembles the bed of a rocky stream, then climb another kilometer up a steep hillside through dust and pine needles and undergrowth, until you marvel at having any breath left at all.

As you ascend, five monarchs flutter across your path, then fifty, then five hundred and then a thousand. By now it is like a heaven, where there is neither thought nor song, only the fluttering of wings. The sun shines in the azure sky and the monarchs dance between the trees and above the trees, down into the valley, to the streams and to the flowers. Then the sun goes behind a cloud and the butterflies (a minute of cooler air lowers their body temperature so they can no longer flutter) rush upward to the peaks. When a hundred of them are flying, they make a serenely beautiful sight. When there are a hundred thousand, there's a murmuring sound in the air, as if one were gently stroking a girl's velvety neck. They rustle and settle into the branches and the bark of the trees and the leaves of the undergrowth; they open their wings, fold their wings, and remain still. They look like angels.

If you lower your eyes to the ground, however, you will see, during the sunlit and the darkened moments, some of the butterflies falling to the ground; dully they wave one wing, rake the earth with two or three legs, and die.

Butterfly death rains from the butterfly heaven. From the point of view of the *Danaus* species, dying is easy amid 25 million. From the point of view of the breathless pilgrim with human habits it is a sorrowful sight, and the pilgrim has a human, though foolish, tendency to warm the fallen angels in his hands, place them back on the branch, try to encourage them to fly. It isn't possible. They have spent their energy sources, used up their fatty reserves. To make up for all the butterflies and caterpillars he has done in, the pilgrim tries to revive the butterflies raining down from the sky. As if with these few victims the secret of migration and orientation in the breeze of Central America would perish.

The beautiful butterfly death rains down from the beautiful butterfly sky, and it is not fitting for humans to interfere. The broken-off wings are steeped in dust, and into this dust the little bodies crumble. William Blake wrote:

The invisible worm
That flies in the night,
In the howling storm,

Has found thy bed
Of crimson joy,
And his dark secret love
Does thy life destroy.

The individual less sensitive than Blake and his descendants will refrain from interfering, because he is not sure how joy may be crimson on a Mexican Oyamel fir tree and because he is sure that this invisible worm of death has been well accounted for. He could even use the fall of dead butterflies as a metaphor to explain to Blakelings (who get excitable about biodiversity and the greenhouse effect) that there is an analogous rain of dead algae *Emiliana huxleyi* from the surface to the bottom of oceans, and that the balance of planetary life depends upon it much more fundamentally than on the fall of brown-orange angels.

On the steep hills of the Sierra and in the valleys where the iridescent turquoise morpho butterflies dally like an earthly luxury, black-eyed Indian children crawl out of their huts, which resemble pigsties with television aerials. Their fathers have lost their jobs in the silver mines in Angangeo and the luckier ones take part in the illegal lumbering of protected fir trees in protected areas. These little children offer glass butterfly pins for sale, holding up in their grubby little hands butterflies that never lived, one butterfly for three new pesos. They will hardly be able to migrate to the north, these children. Instead they will gradually sink to the bottom.

For a long time I observed the little feet of a three-year-old girl child. They were particularly grubby; on each foot a different shoe, worn down unevenly. Hey, that's biodiversity, I said to myself. And I bought myself a butterfly that had never lived, and will therefore never die.

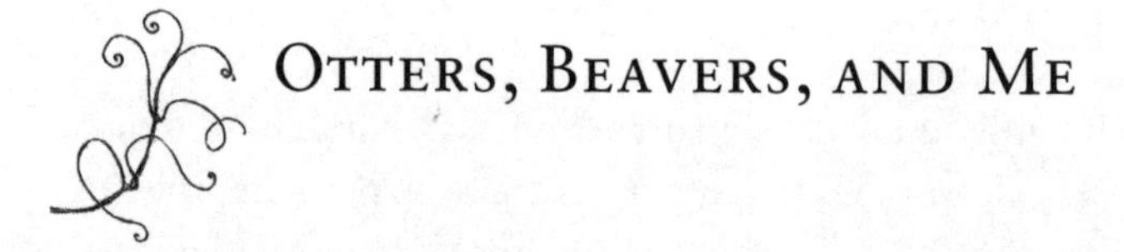

Otters, Beavers, and Me

When I was staying in Tucson, Arizona, I was completely happy. I don't know why it is, but the stony desert, the heat, and the saguaro cacti—which next to our homely cacti are like dinosaurs compared to dispirited green lizards—appeared to me to be a pleasant place to sojourn, even a home of sorts, although I come from the Sumava mountain forests. For me, the Sonora Desert is something like a tonic, giving a feeling of good arrangement and anchorage, despite the fact that the Indians whom you meet on the desert roads, driving their station wagons straight at you, remind you rather strongly that without the assistance of indigenous deities residing on the peaks of Santa Rita, you can take your good feelings and stuff them.

My feelings, however, were affected by Lewis Thomas, who was in Tucson before me, and who remarked in *The Medusa and the Snail* that for a few minutes in this place he lost his scientific distance and delighted in directly observing the beavers and otters kept in the Tucson zoo. Naturally, thanks to the car of a friend who undoubtedly had spiritual connections on the Santa Rita peaks, I hurried to the place; it is called the Arizona-Sonora Desert Museum, but is nevertheless also a zoo.

A deep path cuts through the grounds, and from it beavers can be seen on the right and otters on the left, depending on your direction, first from above, on the bank of a narrow water tank, and then in a tunnel behind a glass wall, from below, hence underwater, and then once more from above. Beavers on the right, otters on the left, or vice versa. Lewis Thomas observed: "Within just a few feet from your face, on either side, beavers and otters are at play, underwater and on the surface, swimming toward your face and then away, more filled with life than any creatures I have ever seen before, in all my days."

Having seen them, he gains an insight about himself:

> I am coded, somehow, for otters and beavers. I exhibit instinctive behavior in their presence, when they are displayed close at hand behind glass, simultaneously below water and at the surface. I have receptors for this display. Beavers and otters possess a "releaser" for me, in the terminology of ethology, and the releasing was my experience. What was released? Behavior. . . . Standing, swiveling flabbergasted, feeling exultation and a rush of friendship.

With Thomas's words at heart, I approached the passage between the otters and the beavers, brimming with the same sense of expectation I might have if a beautiful girl were to emerge from a pool and sit down at my table, or come out of the bathroom in a silky robe.

Alas, I am not Lewis Thomas.

On the left, on a sandy bank, three otters were lying around, obviously bored by the desert museum, the water, the sand, Tucson, the visitors, the fish, and life itself. After a long time, an otter got up, in the manner of a shop assistant who isn't being paid for working overtime, and slipped into the water. But it did not swim toward me; on the contrary, it disappeared behind a stump and later wiggled back to its original place, where it slumped down with a manifest skepticism about an otter future that contained nothing but tree stumps.

There were no beavers on the right, just one, a very obese beaver, lying hidden from sunshine in the tunnel, but out of the water, on concrete, and it was shivering, with its paws turned up. If it did release something in me, it was the memory of a room in a psychiatric ward in Katerinky, where, according to a plaque on the wall, Smetana died, and where deeply depressed patients got electric shock treatments.

Later, talking to experts, I asked why an obese beaver would lie on a concrete platform on its back and shiver. It turned out, however, that zoologists know a lot about beavers but very little about shivering; physiologists know a lot about shivering but next to nothing about beavers; and pathologists refuse to have anything to do with anything that can't be embedded in paraffin.

Following in the footsteps of Lewis Thomas, I did not learn whether I was coded for beavers and otters, and I learned nothing about my distinctive behavior. If I accept his metaphor, then I am fatally related to otters and beavers that are fatigued by the sight of ponds too small to live in and too large to die in.

Ruminating in Tucson the following night, after seeing the film *The Unbearable Lightness of Being*, in terms of those "stereotyped, unalterable patterns of response, ready to be released" with which we are stamped, as Thomas says, I came to the conclusion that my patterns are Czech. Through them, I accept the fact that creatures deemed full of life by an American are apt to be seen as fatally lost by us Czechs, even when we are in an otherwise quite atypical state of bearable lightness of being. It is not just our habit of cyclic heaviness, which probably in itself brings on more cyclic heaviness and which might be infectious and contagious to rodents, even through glass. But maybe Czechs have receptors for creatures in distress, and perhaps some chemical signals for distress imperceptibly guide our steps through zoos and desert museums, and possibly through life. Thus it is that, unmistakably, we come upon otters and beavers at the moment when their habitat is conspicuously reminiscent of the state of the Central European habitat.

While the two types of habitat share a profound similarity, there is not much choice in either of them. In both of them the process of self-realization, the fulfillment of existence by essence, as Mr. Sartre would put it, is rather more limited than it appears to be to people outside the enclosure. In enclosures there are a little less freedom, fewer facts, and a little more communicable suffering. It is not so much that hell is others, but rather that the bars and glass are, for a small nation, both essential and existential. If anything, we are condemned not to freedom but to an enclosed habitat, without which we might not survive anyway. Neither we nor the beavers and otters in the Mojave and Sonora Deserts. The visitor from a different world can feel at moments that both kinds of habitat are full of life and elegant, marvelous movement, just behind the glass, on land and underwater, corresponding to the visitor's receptors for play, creativity, and lightness. He can feel that they make him lose habits of abstraction and facilitate direct observation.

Maybe sometimes we offer just such pleasant sights as the otters and beavers in Tucson did when Lewis Thomas came to see them.

I wish he could have come to Prague.

Shedding Life

A MUSKRAT, also called musquash, or technically *Ondatra zibethica* Linn. 1766—the creature didn't give a hoot about nomenclature—fell into our swimming pool, which was empty except for a puddle of winter water. It huddled in a corner, wild frightened eyes, golden-brown fur, hairless muddied tail. Before I could find instruments suitable for catching and removing muskrats, a passing neighbor (unfamiliar with rodents per se, or even rodents living in Czechoslovakia since 1905), deciding he'd come across a giant rat as bloodthirsty as a tiger and as full of infections as a plague hospital, ran home, got his shotgun, and fired at the muskrat until all that was left was a shapeless soggy ball of fur with webbed hind feet and bared teeth. There was blood all over the sides and bottom of the pool, all over the ball of fur, and the puddle of water was a little red sea. The hunting episode was over, and I was left to cope with the consequences. Humankind can generally be divided into hunters and people who cope with consequences.

I buried the deceased intruder under the spruces in the backyard, and, armed with a bundle of rags, I went to clean up the shooting gallery. The swimming pool doesn't have a drain, so the operation looked more like an exercise in rag technology, chasing the blood north, south, east, west, up and down. Chasing blood around an empty swimming pool is as inspirational as listening to a record of Haydn's *Farewell* Symphony with the needle stuck in a groove. I became very intimate with the blood in that hour, and I began to daydream about it. The blood wasn't just that unpleasant stuff that under proper and normal conditions belonged inside the muskrat. It was the muskrat's secret life forced out. This puddle of red sea was, in fact, a vestige of an ancient Silurian sea. It was kept as an inner environment when life came ashore. Kept so that even though it's changed to a radically different concentration of ions, a different osmotic pressure, and different salts, the old metabolism hasn't needed too much reshuffling.

In any case, the muskrat was cast ashore from its own little red sea. Billions of red blood cells were coagulating and disintegrating, their hemoglobin molecules puzzled as to how and where to pass their four molecules of oxygen.

The blood corpuscles were caught in tender, massive nets of fibers formed from fibrinogen, stimulated by thrombin that was formed from prothrombin. A long sequence of events occurred one after the other in the presence of calcium ions, phospholipids from blood platelets, and thromboplastin, through which the shot arteries were trying to show that the bleeding should be stopped because it was bad for the muskrat (though in the long run it didn't matter). And in the serum around the blood cells, the muskrat's inner-life signals were probably still flickering, dimming, and fading out: instructions form the pituitary gland to the liver and adrenals, from the thyroid gland to all kinds of cells, from the adrenal glands to sugars and salts, from the pancreas to the liver and fat tissues—the dying debate of an organism whose trillions of cells coexist thanks to unified information.

And, especially because of the final chase, the adrenalin and the stress hormone corticotropin were still sounding their alarms. Alarms were rushing to the liver to mobilize sugar reserves; alarms were sounding to distend the coronary and skeletal muscle arteries and make the hair stand up, to dilate the pupils. And all that militant inner tumult had been abandoned by what should obey it. Then there were endorphins, which lessen the pain and anxiety of a warrior's final struggle, and substances to sharpen the memory, because the struggle for life should be remembered well.

So there was this muskrattish courage, and elemental bravery transcending life.

But mainly, among the denaturing proteins and the disintegrating peptide chains, the white blood cells lived, really lived, as anyone knows who has ever peeked into a microscope, or who remembers how live tissue cells were grown from a sausage in a Cambridge laboratory (the sausage having certainly gone through a longer funereal procedure than blood that is freshly shed). There were these shipwrecked white blood cells in the cooling ocean, millions and billions of them on the concrete, on the rags, in the wrung-out murkiness. Bewildered by the unusual temperature and salt concentration, lacking unified signals and gentle ripples of the vascular endothelium, they were nonetheless alive and searching for whatever they were destined to search for. The T lymphocytes were using their receptors to distinguish the muskrat's self-markers from nonself bodies. The B lymphocytes were using their antibody molecules to pick up everything the muskrat had learned about the outer world in the course of its evolution. Plasma cells were dropping antibodies in various places. Phagocytic cells were creeping like amoebas on the bottom of the pool, releasing their digestive granules in an attempt to devour its infinite surface. And here and there a blast cell divided, creating two new, last cells.

In spite of the escalating losses, these huge home-defense battalions were still protecting the muskrat from the sand, cement, lime, cotton, and grass; they recognized, reacted, signaled, immobilized, died to the last unknown soldier in the last battle beneath the banner of an identity already buried under the spruces.

Multicellular life is complicated, as is multicellular death. What is known as the death of an individual and defined as the stoppage of the heart—or, more accurately, as the loss of brain functions—is not, however, the death of the system that guards and assures its individuality. Because of this system's cells—phagocytes and lymphocytes—the muskrat was still, in a sense, running around the pool in search of itself.

Not to mention the possibility that a captured lymphocyte, when exposed to certain viruses or chemicals, readily fuses with a cell of even another species, forgetting about its previous self but retaining in its hybrid state both self and nonself information; it can last more or less forever there, provided the tissue culture is technically sound.

Not to mention the theoretical possibility that the nucleus of any live cell could be inserted into an ovum cell of the same species whose nucleus had been removed, and after implantation into the surrogate mother's uterus, the egg cell would produce new offspring with the genetic makeup of the inserted nucleus.

The shed blood shows that there is not one death, but a whole stream of little deaths of varying degrees and significances. The dark act of the end is as special and prolonged as the dark act of the beginning, when one male and one female cell start the flow of divisions and differentiations of cells and tissues, the activation of some hereditary information and the repression of some other, the billions of cellular origins and endings, arrivals and departures.

So in a way the great observer William Harvey was at least partly right when he called blood the main element of the four basic Greek elements of the world and body. In 1651 he wrote: "We conclude that blood lives of itself and that it depends in no ways upon any parts of the body. Blood is the cause not only of life in general, but also of longer or shorter life, of sleep and waking, of genius, aptitude and strength. It is the first to live and the last to die."

Blood will have its way, I thought, wringing out another rag.

It is the color of blood that makes death so horrible. People and other creatures (unless they happen to be the likes of shark, hyena, or wolf) have a fear of shed blood for this reason. It is a fear that hinders further violence when mere immobility, spiritlessness, and breathlessness can't. A fear that keeps the published photographs of killing or slaughter from being true to life. The human reaction to the color of blood is a faithful counterpart to the microscopic reality, the lethal cascade we provoke by the final shot in the right place. There are an extraordinary number of last things in anyone's bloodbath, including a muskrat's. And if any bit of soul can be found there, there is not one tiny bit of salvation.

They say you can't see into blood. But I think you can, if only through that instinctive fear.

Lucky for the Keres, the goddesses of bloodshed, that no one concerns himself with the microscopy of battlefields; lucky for the living that molecular farewell symphonies can't be heard; lucky for hunters that they don't have to clean up the mess.

Unter den Linden

I WAS STROLLING down Unter den Linden Boulevard in Berlin on a pleasant September morning. A pleasant September morning is fostered by the combination of high pressure over the North Sea with a mixture of summer, Indian summer, the anticipation of autumn, plus time to spare. The linden trees are still assimilating exhaust fumes, and winged insects are still sucking nectar from Coca-Cola cups and puddles of melted ice cream in garbage bins. Today the atmosphere of the boulevard is fostered by the combination of high pressure over Europe with a mix of German imperial edifices, noble statues spiritualized by verdigris, buildings in the international style of industrial elegance, partly under reconstruction, plus organized beggarship.

You could even say that the street has a gradient, sloping down toward the Brandenburg Gate, where one finds the maximum concentration of the worms of decay that appear to have been fostered by the Asian mafia. Under the supervision of the clifflike Lenin on the embassy of the former USSR, swarthy citizens set out a display of goods on the hoods of Mercedes automobiles, on wooden boards, on suitcases and lawns, offering painted wooden dolls, Gorbachevs, Uzbek hats, Russian officers' superhats, military medals and banners. The only merchandise absent are Stalin's mustaches and histological samples of Lenin's brain, but there is a surplus of icons and bayonets. If Marxism-Leninism were liquid, they would sell it in bottles, and if Catherine the Great had worn a bra they could easily sell two thousand pieces of it because tourists crave some real history, never mind that it was manufactured two months ago in a Czech factory. So many bits of the Berlin Wall are for sale in small plastic bags that the wall would have had to reach down to Sicily.

Well, this is the palpable pole of Unter den Linden; here you know what's what: that history is equivalent to a sale of summer socks, that honor is sold for twelve Deutschmarks in the morning and eight at the end of the day, and that good taste is half-price.

Alas, there is the other, shall we say abstract/metaphysical, pole of Unter den Linden, where under the supervision of verdigrised Frederick the Great the lobby of the old library at Humboldt University houses an exhibition called *The Aesthetics of the Abstract: Illustrated Mathematics, Computer Graphics, Mathematical Models*. A modest exhibition, for five days only, not to be compared with the infinite fractal geometry of the hammer and sickle sale.

Of course, not many feet are drawn here—who cares about polyhedrons, trinoids, and minimal surfaces?—but when a foot does cross the threshold, the head and heart say "This is beautiful." But why is it beautiful when it is just the result of an equation, something like $E = [\frac{1}{2} - (7 + \sqrt{1} + 48 g)]$, where E has to be an integral, E is the number of peaks, and g is the polyhedron genus, that is, the number of tunnels passing through it? Why is it amazingly beautiful when a mathematician just sucked it out of his pencil as a discrete version of surfaces with constant negative curvature or as an equivariant Willmore surface?

What we see, in azure and pink, is a druse (a crystalline

surface) of something resembling snails, swollen at the periphery and thinned out inside, with tiny channels winding on and on; a biologist is reminded of the inner ear multiplied by three, or the shell of the nautilus; it is as elegant as the firebird's flight, and yet it originated in a conformal transformation of space, or translation of tension, rotation, or mirroring on spherical surfaces.

On the table stand solidly materialized knots, like the twisted plump knots Henry Moore used to make, and they turn out to be 3-spheres, models of a geometrical object in four- dimensional space (except that only a 2-sphere in three-dimensional space is feasible and reasonable for a brain raised on simple arithmetic). A 3-sphere, however, brings us closer to understanding the universe, while a 2-sphere successfully prevents us from doing so . . .

Next to this exhibit, Moebius strips with their three holes are coiling and curling modestly, though terrifyingly, into ear-lobes and shapes of the kind that entranced Salvador Dalí, Georgia O'Keeffe, and the possible designer of our pelvic bones. I was delighted to learn that I could have Moebius strips in the privacy of my home (I've always felt that the essence of home is in small whimsical objects and toys, like boxes, butterfly stickers, and miniature landscapes, possibly with railways). In privacy I twist paper and glue it together, and I am supposed to discover that it is a one-sided surface and that I will achieve mathematical enlightenment if I cut the tape in the middle. I believe this is my only chance to escape the line of street vendors selling 2-spheres in which snow keeps falling on Yeltsin's roof or on Saint Vladimir's Cathedral.

On nearby panels roam pavements composed of one or more types of stones or elements, periodic and nonperiodic, possibly Penrose pavements, as well as the nonperiodic repetitive pavements, possibly three-dimensional, that Paul Klee and

Vasarely imagined, those that constitute the unlikely basis of all op art and design: the substances from which are spun the fabrics of fashionable coats and jackets.

In the back, videocassettes show minimal surfaces derived from soap membranes on wire loops; molecular rainbows shimmer on glittering silvery surfaces and the video voice-over intrepidly points out constructions where the stress is identical at all points, which leads to the derivation of tents and roofs and stadiums. Strategically applied pins shrink the soap membranes into lines connecting several points and produce a useful communication network, one that would be appreciated even by a street vendor if he were told about it by the boss with dark glasses. But it has already been appreciated in practical terms by Moholy-Nagy, Picabia, and Frank Stella.

Unattractive paper dodecahedrons and icosahedrons hang overhead. These, however, have given rise to the industry of Christmas decorations and children's school handicrafts, pleasing even street-vendor parents. Let us admit, though, that even the good Lord would have found the term *icosahedron* obnoxious.

The funnels, denoted by experts as catenoids, are derived from two-hundred-year-old mathematical problems; their mushroomlike associations remind a biologist of tapeworm heads or large viruses in a scanning electron microscope. Jean Arp, of course, would have been reminded of Jean Arp. Formations produced by projecting infinite hyperbolic space inside a sphere with a hyperbolic octahedron whose six top points are placed in the infinite space are strangely close to the three-dimensional drawings of Escher.

There is a timid reminder of the paradisic fractal structures in the deterministic chaos of nonlinear systems—a lovely term for these computer results arising from the algorithms and principles of feedback. The deterministic chaos of nonlinear systems produces results of universally acknowledged beauty;

they are in fact the only artifacts that might adorn both a lacemaker's corner and the study of a physicist, geologist, or botanist; a cathedral of the brave Saint Stephan as well as the pub where the brave soldier Schweik used to go for a beer. That chaos is, at the moment, the source of the most universal images and suggests that algorithms, feedback, and computer projections are, unfortunately, a part of what is often called the soul.

If the Stoics could handle the mundane troubles stemming from transformations of market mechanism and the democratic division of competencies so well in their time, it must have been due to their metaphysics, promising after-death contemplation of the movements of the stars—unearthly beauty, as Ionesco wrote in his diaries.

I am afraid that the new mathematics brings that beauty down to the here and now. And I'm also afraid that if absolute reality is impossible for the simple mind to know and capture, impossible for customary notions to cover, and accessible only to direct experience, then Buddhists and their followers are in the same boat as the children of the hellish new mathematics: the experience of the highest spiritual principle can be achieved only by meditation or by solving an equation.

Galileo must have known this. "The Book [of the Universe] is written in a mathematical language, having triangles, circles and other geometrical forms for symbols, without which we would be speechless and vainly wandering through a maze." Thales, the first geometer, must already have recognized, however, that understanding cannot be derived from observing the geometry of nature, but must come from what humans put into the spatial relationships by their conception of them. That is, from a priori ideas. The salvation of the general validity of knowledge, including mathematics, depends—a priori—on the existence of synthetic judgments, Kant believed. That is, on the agreement of the forms of thoughts with the forms of structures, mass, universe, organisms, communities.

We've gone quite far beyond Galileo's circles and triangles, but the latent common language is still there. If "the best science comes from a combination of analytical mind and esthetic sensibility," as R. S. Root-Bernstein wrote, then this common language is intuition, and it is occasionally used in science to take a shortcut.

And so my aesthetic participation in mathematical disciplines quite refreshed me that September day on Unter den Linden. As a matter of fact, I could claim a number of stony gentlemen—for instance, Messrs. Leibniz, Fichte, and von Helmholtz—on the pedestal of the green-tinted Frederick the Great as kindred minds. High pressure was passing over the North Sea, and the "Hero of the USSR" medal was still selling for ten Deutschmarks at the Brandenburg Gate.

The gradient of Under den Linden Boulevard now acquired a new meaning. For when they are all forgotten, all those fold-up Brezhnevs, all those stolen assault knives, all the pathetic little bones of the German and Russian empires, when a biotronic vivarium replaces the Reichstag and dolphins swim through the Brandenburg Gate, there will still be trinoids, minimal surfaces, and fractals; then, when not even a single hair of Stalin is left, all of Lobachevsky will still be there, and radiantly beautiful.

The Discovery: An Autopsy

Today I had a frightening dream. I was scheduled to deliver a lecture on a scientific subject. The audience was already seated and waiting, but I had only two sets of four or five slides, both of poor quality and unrelated to the topic. I tried frantically to decipher them, in vain. And I woke with a sense of crushing loss. It was like missing a loved one or forgetting the key to some important solution.

What is so emotional about science?

Almost everything. Most notably success and discovery, whatever they may be.

The concept of discovery has always been treated with exceptional thrift in the sciences. And sheepishly. It is a word that points to the past, something over and done with, like a little grave decorated with a philosopher's stone or a medal from a long-forgotten war. In addition, there's something trivial about the concept of discovery because we do it every day, discovering lost keys, mislaid notes, an odd sock, the first swallows. Speleologists discover new caves by rolling a boulder aside. Schliemann discovered Troy. Livingston discovered Victoria Falls. Halley discovered Halley's comet. N. R. Grist discovered the nude mouse mutant, Huntington the choreic disorder of the central nervous system, and Hassal the thymic corpuscles. Really good, tough discoverers and explorers are supposed to sacrifice their lives and matrimonial happiness, be afflicted by the diseases they describe, or at least be beset by mortal fatigue and nervous exhaustion, in proper geographical locations, among avalanches, tsetse flies, natives, and superstitions; if they enter the world of legend, they are set upon by tempters, bloodsuckers, devils, and many other asocial and, yes, even undemocratic forces. Remember Faust.

Discoverers frequently perch on the shoulders of other discoverers, quoting from them and thereby risking, on the one hand, falling with gravitational acceleration and, on the other, a considerable blurring of the image of when and where a certain thing was discovered or, on occasion, when the discovered was rediscovered. Even the aphorism of Diego de Estella—dwarfs on the shoulders of giants see further than the giants themselves—published in 1622 has been quoted and repeated so many times since that it is attributed these days to about thirty other people.

The phenomenon of discovery and scientific creativity can appear anywhere, apart from scientific monographs and scientific communications, and even apart from the memories of active scientists.

The discovery contains within it a certain essential immediacy; it occurs "now or never," or once in a great while when "if God allows it, even a hoe goes off like a rifle," as the Czech proverb says.

Newton, however, in his time, in answer to the question of how he discovered the law of gravity, answered: "By thinking of it continually."

The discovery of the DNA model, the subsequent scientific revolution, and the beginnings of molecular biology in 1953 seem less of a discovery, a revolution or even a new beginning when one reads Watson's 1968 book: no romance, no heroism, no enlightenment, only the MRC laboratory in Cambridge, many open discussions in which people said anything that occurred to them—call it brainstorming—plus knowledge of the literature and great familiarity with the results of others. Watson's book, *The Double Helix,* is a "candid portrait of the scientist as a young man in a hurry," as Robert Merton appositely wrote.

Little can be learned about discoveries from contemporary scientific media. First of all, in experimental and theoretical sciences, discoveries become obvious only after a few years. Second, communication is usually based on subjective simplification that omits all the jumps of thinking and fits of inspiration. This is necessitated by the imperative of brevity and condensation on the one hand and, paradoxically, by an attempt to sound objective on the other.

Thus Frederic Holmes, in his studies of scientific creativity and its history, discovered that it does not suffice simply to read a researcher's technical papers. It is necessary to go into his protocols and, if possible, to speak with him to acquire a complete picture. To look through the keyhole, as Peter Medawar wrote. A computer program based on Holmes's study of the biochemist Hans Krebs establishes that the discovery of the ornithine cycle was the result of a "whole sequence of tentative decisions and their consequent findings, not . . . a single 'flash of insight,' that is, an unmotivated leap," wrote Kulkarni and Simon.

The concept of discovery often implies something referred to as intuition; according to Henri Bergson, it is an instinct that has become disinterested, self-conscious, capable of reflecting on its object and enlarging it indefinitely. Even though instincts and intuition are more agreeable to publicists than to psychologists, everyone (from publicists and psychologists to physicists) can remember moments when something has occurred completely unexpectedly, when "it" has come from out of the blue. One forgets that things come to us "out of the blue" more frequently when we have previously been overcast with heavy clouds, toiling with something, or when we have arrived at a point in our lives where we have acquired enough fundamental and extensive instructions—in other words, school education, or what's left when we've forgotten the majority of what we've learned. In so-called intuition nothing brand-new jumps in. Something emerges that has been covered over and has remained beneath the surface, beneath consciousness. Ideas, optimistically described as intuitive, come to us as a rule in the early hours, when our batteries have been recharged and day breaks under our sleeping caps. Milton Rothman compares intuition to a computer whose central processor is working, but nothing is showing up on the screen. Finally, when the process is complete and somebody switches on the monitor, one becomes aware that the result has reached a conscious level as something new, like a spark, a star, like a discovery. Only detailed mental archaeology can help us (internally, privately) to explain that it was not a case of something from nothing.

Even in science, innovation is usually preceded by imitation, as J. Sasso remarked in "The Stages of the Creative Process."

In fact, no idea from an idle mind can replace the work carried out by a working hand. "Propositions arrived at by purely logical means are completely empty as regards reality. . . . All knowledge of reality starts with experience and ends in it," said Einstein in the Herbert Spencer lecture in 1933.

Small wonder that the American scientific historian is shy not only of "discovery" but also of "creativity" in science and infers that it may be more appropriate to replace the much-abused term *creativity* with some less pretentious word, such as *style*, as the biochemist J. S. Fruton wrote.

In addition, any creativity and any style imply a genius loci, or a stupor loci. Our Czech genius loci would say, I may have made some sort of discovery, but I'm not sure. Our genius loci instructs most people to do "normal science," which is a kind of puzzle-solving (Thomas Kuhn's term); scientists apply current theories as the rules of the game, instead of attacking the leading paradigms. We like peaceful science.

I recall a project that led to the publication of my most quoted work of the years 1962–65. It was on the subject of why we have lymphocytes, those billions of white cells in the blood and in the lymph nodes, spleen, lungs, intestinal wall, thymus gland, lymph, and everywhere in the various interstitial tissues, those billions of vagrant lymphocytes, a total of one and a half kilograms per human body. It was evident that they had a role to play in inflammation and in the defense of the body, and that they belonged in the world of immunology. But it was not at all clear how. Immunology was, at that time, particularly concerned with the formation of antibodies, and antibodies, according to the dominant hypotheses, were formed in plasma cells differentiating from somewhat obscure large reticular cells of connective and lymphoid tissues.

It was simple to set the lymphocytes with the antigen into a tissue culture, but we did not have much success with them; to get the lymphocytes without the admixture of other cells was difficult, and to get certain chemicals proved to be a Sisyphean task. When I ordered ACTH in 1956, the order ran several courses around heaven and hell, but absolutely nothing came of it. I completed my thesis without ACTH. After I submitted it, however, ACTH not only arrived but continued to arrive, so that, over the years, there was enough for a small hospital. Some of it was past its expiration date, and no one knew what to do with it. No one knew how to stop the showers of ACTH ampules. It was called planned economy.

By chance I came across the literature of the technology of diffusion chambers, small capsules or cylinders created from two membranes that were impermeable for cells but permeable for proteins. The cells would be closed inside with an antigen and chambers implanted into the abdominal cavities of the recipient, which would provide nutrients by peritoneal exudation diffusing into the chambers. But the recipient couldn't attack the cells inside if one was dealing with an animal of the same species. We used baby rabbits (which did not react, or barely reacted, to the antigens we used).

When we were inspecting the rabbits we found a small sac in the dorsal abdominal wall, behind the left kidney, a cisterna chyli from which we could easily and quickly collect lymph containing several million pure lymphocytes, more precisely, lymphocytes looking like lymphocytes. These we mixed with antigen and put into the chambers, then put the chambers into the baby rabbits, a process that did not bother them in the least. It is better for a baby rabbit to have a sterile diffusion chamber than malevolent coccidia in its belly.

There wasn't an ounce of intuition in this unless you consider poking around in a rabbit's abdominal cavity intuitive. There was a lot of patience involved, some imitation, and variation on the experimental design of our "Prague school" of the time. The "style" was "poor people cook from water" with a tiny admixture of heresy, since some bosses still subscribed to the Russian "nervism," which assumes that nerves are involved even in antigen recognition and in the production of lymphocytes from acellular matter: beliefs somewhere in the neighborhood of Needham's production of lice from a dirty shirt.

The so-called discovery occurred when we began to take out the chambers with the lymphocytes after eight, ten, and fourteen days of cultivation inside the rabbits. We discovered that the lymphocytes had changed to cells producing antibodies which could be visualized by immunofluorescence and even cells that were likely to present antigen. In other words, the lymphocyte was the cell in which immune reactions could be triggered and completed.

It went against everything we had read in books.

I don't even remember if the discovery brought joy. The feeling was defined by Picasso's adage: The *against* is prior to the *for.* The first and characteristically Czech reaction was not "Eureka!" but "Jesus!" The result said yes to the working hypothesis, but we are usually alarmed when results say yes. More generally, we are alarmed by clear results. In most of our history, results are ambiguous. Clear results make demands. They demand clear and firm responsibility. They demand a high standard of quality at every step. Well, the chambers were made from U.S. Millipore membranes, but what about the sealing? So we repeated everything using new glues and polymer coating. Same results. Is a Czech rabbit as good as an English rabbit? Didn't it catch some lunacy from the Russian rabbits (which, according to Russian references, would support independent differentiation of Russian lymphocytes without any diffusion chamber protection)? We used Belgian rabbits; same results. But was even the Czech lymphocyte the real international lymphocyte? If so, how come nobody had noticed its potential over the last ten years? The Czech discovery is immediately linked to Czech feelings of guilt and error.

The detailed and protracted observation of lymphocytes in diffusion chambers—and wherever else in the world—naturally creates a certain personal bond, a feeling that cannot be easily defined, a bond with the lymphocyte, with its ability to transform itself, and with its limitations. Every conscientious and honorable tradesman has a feeling for his materials. Every pathologist must have a feeling for the material seen through the microscope so that he doesn't confuse chronic gastritis with sarcoma (as happened frequently in the case of a self-appointed Czech "pathologist" who established pathology on a new Marxist-Leninist basis because it had previously been the domain of conservative truthfulness and honesty, and those didn't appeal to a "reformer"). All good clinicians develop a sense for the signs of illness and health, deviations and norms, and it depends on the measure of their imagination and the extent of their education how skillfully and proficiently they use this sense. This sense is related to intuition in the way a sparrow is related to the phoenix.

If one were merely discussing biology and medical practices, a researcher would be suspected of replacing an unscientific feel for the material for the scientific method of greater or more exacting disciplines. Even the greatest moguls of hard science, however, consider this sense to be a fundamental step toward discovery.

Barbara McClintock (Nobel Prize for the discovery of new genetic processes in plant life) is indebted to her developed "sense of the organism." Joshua Lederberg (Nobel Prize for his work on the recombination of genetic material and its organization in bacteria) talks about the necessity of imagining oneself within the research situation: "What would it be like if I myself were a bit of DNA in the chromosomes . . ." Santiago Ramón y Cajal (Nobel Prize for his work on the structures of the nervous system) operated according to Sherrington's testimony about the microscope as a miniature world in whose field of vision the little creatures and subcreatures live, struggle, love, yearn, and sacrifice themselves as humans do.

Even the physicist and theorist of science Michael Polanyi judged that understanding begins with "personal knowledge," a mental identification with the subject of study, a transference into it. The metallurgist C. S. Smith, who worked in Los Alamos during World War II, stated that, during his work, he had the feeling that he himself was the alloy and he acted as *it* would have done. The terminology of science seemed to him second-rate: the moment of discovery was absolutely sensuous, and he needed mathematics only for communication with others. Even the mathematician S. M. Ulam sometimes tried not to count numbers or symbols but the almost tangible impressions linked with logical proofs. Carl Friedrich Gauss, asked how he had come to his theorems, replied, "Through systematic feeling around."

Hannes Alfvén (Nobel Prize for the physics of plasma) did not deliberate with the aid of equations. He was aided instead by imaging what the world is like from the point of view of the electron and which forces drive it to the left or to the right. And Einstein, in his musings on gravitation, imagined the subjectivity of light rays flickering across the walls of a falling lift. He tried to visualize the ray that would follow the speed of light. He wrote: "The psychological entities which are the elements of our thinking are certain signs and clear or less clear outlines which could be freely reproduced and combined."

Feeling for a thing, personal knowledge, and the ability to identify with the research subject attempt to fuse the sensuous experience and the scientific imagination linked to it. In a way, the process is a counterpart of philosophical insight *(Anschauung)*. Metaphorically, it is a kind of brain economy, the right hemisphere helping the overburdened left hemisphere. In athletic terms, it is better coordination of the body, of the legs and hands in a cross-country run, not a shortcut that would illegally reduce the distance.

In the scientific process, the feeling is important not so much for understanding, but for decision making. In my lymphocyte story, it played a role from beginning to end; it was different from intuition simply by virtue of the fact that it was carried out over a long period of time, in repetitive cycles, there and back. It was self-critical, like a back projection and like a counterpoint system of logical, verbal, differently formalized or abstract considerations. It helped me to forget the technical worry beads and to concentrate on one definite interpretation of the observation.

I can imagine that in less sophisticated scientific disciplines, in the good old times, a proud independence and one-against-all attitude was possible. In dynamic modern sciences, where a total overview of all literature and all possible implications of a finding are beyond individual limits, some kind of endorsement, some sense of solidarity, and, always, some discussion in the closest scientific circle are inevitable. In our particular case, it was also a solidarity of literacy against the forces of bureaucratic illiteracy asking questions like "What is it good for?" and "Why don't they quote from the great Soviet science?"

This solidarity was a source of some inner satisfaction and made it possible to enjoy the whole process as a boyish adventure. Which is, of course, the best part.

The second part is publication, in the shortest available time, following Parkinson's First Law (work expands to fill the time available for it), and in the best possible journal and in the clearest possible form, following Parkinson's Third Law (expansion means complexity, and complexity decay).

This part of the business is again closely attached to the genius loci. There are—and always have been—suspect geniuses and reliable geniuses. The same finding—or discovery—looks much better when it comes from New York University than from Eastern Europe, and the gap is growing. That time, I got the paper into the *Annals of the New York Academy of Sciences,* after a short report in *Nature,* and it appeared almost simultaneously with a paper from Oxford, reporting the same truth about the lymphocyte. Even a discovery becomes a discovery only when everyone else has also seen it. Seeing something alone is only for visionaries, fanatics, and frauds.

As a souvenir of the tiny victory, I've kept two boxes of slides recording the story of the lymphocyte, and all the Millipore and Plexiglas materials in the lab. With all my books, poems, metaphors, and lines, quoted and disseminated in the mind of the literary public, I still feel a lot safer in the memory network of science. In literature, we are miniature giants. In science, one can become, once or twice in life, a middle-sized dwarf.

And, as is well known from fairy tales, dwarfs have many more emotions than giants.

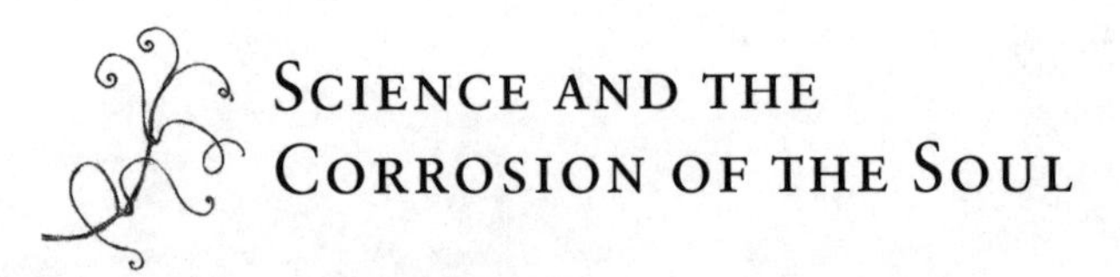

Science and the Corrosion of the Soul

Bryan Appleyard says quite clearly in his 1992 book *Understanding the Present* that science affronts human dignity: "Science is not a neutral or innocent commodity. . . . Rather, it is spiritually corrosive, burning away ancient authorities and traditions."

Appleyard believes that science has developed since Galileo into an autonomous entity, with a life of its own, progressively detrimental to human existence. No wonder, then, that *Understanding the Present* has been quoted in editorials in scientific magazines such as *Nature;* a fundamental misunderstanding of science rarely gets expressed so vigorously—the tone makes one think of parliamentary debates—and seldom achieves such commercial success.

Appleyard, echoing Francis Fukuyama's *The End of History,* thinks that science endangers human survival on the planet despite the positive role it has played, together with technology, in the victory of liberal democracy. Liberal democracy is no triumph either, since it allows too many things that are adverse to what they call "human essence": embryo research, abortion, animal experimentation, environmental destruction, development of uncreated species and unprecedented polymers, nuclear technologies, the human genome map, and other unnatural activities.

At this point, someone should define what would constitute a human triumph in the antiscience conception, but the critics usually don't get around to that.

Of course, the issue is the "natural world" of sense and meaning in which we live as concrete and more or less replicable beings. The difficulty with the natural world is that it is being addressed and handled both by scholars, competent philosophers of the Husserl camp, and by simple minds of the New Age style, for whom Husserl's book on the crisis of Western science is the same kind of trap as molecular genetics. Where the philosopher whispers, the simple mind shouts.

A second difficulty lies in the fact that at the individual level of a thinking and feeling man or woman, the natural world is supposed to be grasped by the human mind as if the mind itself were a thing apart; this is logically impossible. The human cognitive apparatus is a product of biological evolution, and its mission is to guarantee the survival of the species, not "recognition of truth," as the Slovakian biochemist and philosopher Ladislav Kovac has recently pointed out. The recognition of truth can be a recognition of the aspiration toward a supraindividual cognitive system. It's a wonderful idea to imagine somebody or something evolving to the state of real or final wisdom, wisdom for all of us. So far, this hasn't happened to any individual soul or through any spiritual movement. So far, it has only been the trend of the cognitive networks of modern sciences. They are both objectively and subjectively "the reason which owns us," in F. W. Schelling's words.

More surprisingly, in all such argumentation, we never really get a definition of the "human essence." If it is something unchanging, at the core of each man, woman, and child, why get upset with animal experimentation, since we have always survived at someone else's expense and "natural" cultures have so often performed bloody human and animal sacrifice compared to which a surgical intervention on an anesthetized animal is a pastoral procedure? If there was something invariant in us we certainly became alienated from it at the crucial point of self-reflection, in our relationship to death, and at the very beginnings of the cultures and civilizations called "European."

If, on the other hand, this "human essence" is a changing quality, which is the basis of my faith and hope, then changes must also occur in the content of expressions like "human," "being human," "human dignity," and "humanitarian approach." Then there is no point in regretting that ancient authorities and traditions get burned away. It becomes hard to put any trust in the urge to return to fundamentalisms, from religious ones to the ecological fondness for all things visible and sentimental as opposed to things invisible and outside the range of our feelings. We must learn to ask, on a case by case basis, whether a proposed view is a relapse to a more primitive stage of the spirit, soul, or mass mentality, or whether it is a response not only to civilization's problems and to scientific viewpoints, but also to Westernization, which for many societies of the Third World is synonymous not only with economic growth but with survival itself.

In *The End of History and the Last Man*, Francis Fukuyama contends that the historical evolution of liberal democracy is a proof in itself of how human nature has changed over the last couple of millennia. Heidegger took Nietzsche's radical historicism to the conclusion that any traditional ethics or morality sooner or later becomes impossible. In terms of approach and method, Picasso said: "We always stick with the old-fashioned ideas, with outdated definitions, as if it were not the very task of the artist to find new ones." And he was well aware of what he was burning away. But what is all right when it is said by a historian, a philosopher, or an artist becomes outrageous when it is uttered by a geneticist or a physicist.

Ultimately, then, how can we believe that science is an autonomous, aggressive, and oppressive force or power? Are we saying that it's something like a product of mutants from dimension X? The essence of radical historicism is derived from science, and science is, I believe, as integral a part of modern humanity as art (though I won't bet, even if just from professional shyness, as Richard Rorty does, that poetry's evolutionary chances are better than philosophy's). Science is part of our spiritual climate, just as that climate is the consequence of our activities, including scientific ones.

The increase in scientific knowledge, wrote Mario Vargas Llosa, undoubtedly influences history. However, there is no possibility of rationally predicting the development of scientific knowledge.

Was anything wrong with the Viennese atmosphere of the early decades of this century, from which emerged not only new disciplines dealing with the subconscious, with sexuality, and with language, but also new approaches to the demarcation of sciences and pseudosciences? Why would Freud be less corrosive for traditional human values than Karl Popper and Ernst Mach? Doesn't the cultural atmosphere of Vienna, or at other times the climate of Copenhagen, Cambridge, or even Prague, show that science is not an autonomous entity but a natural component of a creative atmosphere and of human progress, or at least of the process of solving the soluble?

Either we respect even what we do not understand or we pretend to understand only what we respect. Unfortunately, respect is not very closely connected to knowledge. Sometimes it is entirely disconnected.

I admit that these are the opinions of somebody who has been trying to do both science and art for a long time. Maybe my soul has been corroded, or it has refused to be inflated by the hot air of group mentality. Maybe my soul has been lost altogether. But I know at least a few artists with an obviously healthy soul who respect science as a way of acquiring knowledge and as an integral part of Western culture.

What corrosive forces or what supportive factors have I experienced thanks to scientific work, at least in terms of self-reflection, and, if possible, without superstitions and illusions? The main limitation I'm aware of is that I'm unable to accept any other mode of acquiring knowledge about the world of nonself than the scientific mode, the one that is acceptable for professional scientific criteria in various disciplines, the one about which I know, from my own experience and from literature.

In other words, I am unable, in principle, to share the axioms—I would call them myths—Milton Rothman exposes in his book *The Science Gap* (1992):

> Nothing is known for sure.
> Nothing is impossible.
> Whatever we think we know now is likely to be overturned in the future.
> All theories are equal.
> Scientists create theories by intuition.
> Advanced civilizations will possess forces unknown to us.
> Etcetera

Neither can I accept the traditional poetic myth that "the scientist doubts, the poet knows," since I know from everyday life and work that poetic knowledge outside a poem is not worth a wooden nickel, and any other teaching of the men of spirit or the men of practice will not last very long, with the exception of several scientific and technical fundamentals with everyday application. In most situations we are like the courtiers from the poem "Seekers after Truth" by Dannie Abse:

> Below, distant, the roaring courtiers
> rise to their feet—less shocked than irate.
> Salome has dropped the seventh veil
> and they've discovered there are eight.

The consequence of the above-mentioned scientific limit is, however, that I am not alone with my private history, my own mental restrictions, my stupidity and ingenuity, my inventiveness and forgetfulness; I share the notions (knowledge and know-hows, as Vilém Laufberger used to say), gained by observation, experiments, judgment, and computations by other trustworthy persons who convince me, correct and complement me, and mostly surpass me by far, which is a pleasant feeling for anyone who does not see himself as Prometheus.

Another limitation is the loss of "the pleasure of transcendence," or more specifically the pleasure of personal transcendence. This pleasure, as Harold Bloom writes in his book *Agon: Towards a Theory of Revisionism* (1982), is "equivalent to narcissistic freedom, freedom in the shape of that wilderness that Freud dubbed 'the omnipotence of thought,' the greatest of all narcissistic illusions."

It is of interest that the omnipotence of thought is demonstrated by predisposed individuals much more readily in the sphere of macrocosms and microcosms than in the sphere of communal hygiene or local administration. In politics, the frontiers of free invention are surprisingly much more tangible than in the treatment of rheumatism and in time-space theory. We limited individuals, less disposed to free invention, understand that the earth is not flat, that the center of the universe is not in our solar system, that Newton with his laws of gravity was more reliable than theoreticians of UFOs, that photons are unchanging and permanent entities, that one cannot exceed the speed of light, that life is based exclusively on biochemical principles such as the self-perpetuating information systems of nucleic acids, that even spiritual existence is based on impenetrable complexity and on the order-out-of-chaos principle, that a swan cannot be crossed either with Lohengrin or with Leda, and that there are only two surpassing, transcendental categories in life: the genome and the extracorporeal heritage that is sometimes called culture and sometimes civilization.

There is, however, the inner circle, the dark world of the Self, located within reach of our knowledge, know-hows, and decision making in the same way as the secretions of glucocorticoids and the reserve of stem cells in the bone marrow are. That is, it is located within reach, but we are not very good at reaching it, although we need not make a virtue out of our difficulty. The darkness of this inner world is demonstrated by the fact that the description Lewis Thomas offered in his 1974 essay on the autonomy of organs also applies to it: "Nothing would save me and my liver if I were in charge. . . . I am . . . constitutionally unable to make hepatic decisions." It is also demonstrated by the fact that, as Thomas says, there are quite a number of selves, and they come one after the other, with others waiting in line for their turn to perform, and the worst moment comes when one wants to be just a single self. Some manage it—at least Thomas does—only when they are listening to music.

As I know from my own experience, there is very little of that singular Self in the dark inner world, but there are a lot of events and images from the outside. A soul expert calls them archetypes and makes them part of some tribal subconscious, when in fact what I have in there is this year's kangaroo from Wilpena Pound in Australia, the frightened face of an elderly lady across the aisle on the plane, tumors in the satin lung tissue of an unknown young man I dreamed about last night, the hardly audible violin music in the garden during sunset yesterday evening, the outline of a girl I once loved, and the figure of my mother, the quiet voice of a girl without a father, who cut her face with a kitchen knife—it's nothing, she says, you just press it to your cheek and slice—obstacles to the publication of this scientific study, ways of surmounting them, the voice of Dana H., melodiously interpreting a film on the sexual habits of ground squirrels . . . In fact, I even have in there—and I don't regret it—the paradox of Schrödinger's cat, several NK cells, a couple of nude mice, the anthropic principle of the universe with Doctor Grygar's profile, and the TNT mice probe with the dark eyes of Doctor Marie L., four unanswered letters, a mess involving my phone bills, the first eight lines of the *Iliad,* and a spot on my beige jacket.

I would not compare my inner world with the transcendent soul, the hypertrophic soul (in Milan Kundera's term), but I'm guessing that each of us has a soul just the right size.

I do not know if my personal myths, feelings, and illogical urges are good for anything. They are here, they are strictly personal, they do for a sort of poetic attitude. They are simply, in the literary application, private idiosyncratic metaphors, symbiotic plants in the "secret gardens of self," in William Carlos Williams's term.

In view of these secret gardens I have no doubt about the role of poetry (in the broadest sense of the word) in my life, in every life, in the average human inner world governed by principles of uncertainty. I do need poetry as a sort of consolation, a temporary relief and limited hope about my personal future. I do not need poetry or religion to explain anything about Self and non-Self. I need poetry as the last possibility for saying something against gravitation, against the degeneration of nucleic acids and arthrosis, although I know that it does no good.

Limitation as recognized by scientists (as opposed to poets, fundamentalists, New Agers, and other playful minds) is the awareness of what is possible, of the borderline between anarchy of thought and scientific courage.

Ultimately, science limits the spirit (or soul) in the technical sphere. When you are doing something in the laboratory, something mechanical like weighing, or something magical like seeking fluorescent cells in black darkness, you are outside the inspirational sources of art or any other inventiveness and creativity. Your mental activity is focused on digital counting or maneuvering in the dark. You are totally devoted to figures or to darkness. If you do it forty hours a week, there is not much time left for creative mental activity. A soul craving other levels of experience must learn patience and humility. Given those constraints, writing poetry or anything else becomes a kind of recreation. Such a view may appear to be a terrible limitation to some artists, but it does not appear so terrible to me, especially when I consider how many artists have sold their talents for commercial success, how many former researchers have embarked upon careers as entrepreneurs, moneymakers, or, most horrifying of all, political big shots.

Science teaches us some discipline. It forces us to concentrate on a single problem and a single approach to it at the given moment: only then is there hope of achieving some tiny success in any research, including art. In a purposeful activity there is no room for free floating. Scientific practice has taught me: That there is a big difference between real involvement of the mind and hand and a purely internal "creative" thought. That lengthy social exchange of monologues based on "what I am" cannot replace "what I have done." That ten minutes are ten minutes even at a lecture or a poetry reading. And finally, that criticism is not sublimated violence or a plot by depraved individuals to destroy someone's progress; very often it is a qualified assessment of our work by others, naturally in the conditions of the existing climate or paradigm. Or dogma.

There's a moral component to be derived from science, and

most humanists don't like this at all. Alas, most humanists are defined mainly by never having worked in science. They have never read Jacob Bronowski saying that science nowadays requires rigorous thinking and a scrupulous code of conduct in the laboratory, library, or operating theater or at the control panel. A rigorous code of conduct is not very satisfying in terms of the free human soul, but the soul does not have much fun even during the daily routine, for instance, when replacing a fuse or waiting at a traffic light. Today, the scientific mode of conduct and inquiry presents a model of behavior that will gradually become more and more necessary, even in the nonscientific sectors of our dense civilization's time-space.

I don't deny that one can develop habits of rigor in the pursuit of the arts and humanities. But in science and technology you acquire an exacting attitude more quickly, and you learn that deviations will be punished. We know of many cases of fraud, error, and human failure in science and technology, from plagiarism to experiments like Chernobyl. By contrast, I'm not aware of many cases of fraud and failure in the arts, maybe because they are noticed rarely and by very few people.

Sadly, we no longer learn information on the norms of life from philosophy and from poetry; we learn them from technical instructions, which have a linguistic and aesthetic quality that approximates the level of communication among social insects. Their influence on the soul is undoubtedly negligible, but it's better to have some norms than none at all. Science is not to blame; the norms reveal not so much the shortcomings of science as the shortage of it.

And yet, the soul—that dark blossom in a secret garden—does have an impact on science itself. It too is an integral part of the existing atmosphere and genius loci, which I have eulogized elsewhere. Public and private metaphors, fantasies, civilized myths, or myths favored by common sense—all these affect the time and the place, and through them the way we are tuned and our half-conscious affinities for judgments, hypotheses, estimates of probable and possible events. They have contributed to the fact that the evolution of science cannot be assessed by rational methods.

It is something like the zeitgeist, composed of little individual idiosyncrasies and group mentalities and paradigms. "The objective spirit," says Norbert Bischof in a 1993 book on Konrad Lorenz, "has a human dimension in addition to the material dimension, and it has always been known. But we did not like psychology to get too close to the human factor. . . . When searching for hidden motifs of creativity, they were evaluated in terms of morality instead of being analyzed in terms of causality."

If it were true that "the reason is basically just the effector of our emotions," as Ladislav Kovac writes, then the collective mind and reason might be one of the few ways out of the comical state in which the individual reason is able "to transform any common hormonal disorder into a metaphysical concept."

However, it is not desirable for souls or zeitgeists to undermine the self-confidence of entire scientific disciplines, or to influence collective processes of questioning and procedures that lead from the insoluble to the soluble. Why relativize those few scientific certainties we have, and force researchers to apologize for being here and for having achieved something?

I would like to share Mr. Appleyard's complaint, but I have nothing to complain about. Science has taught me to say "although" and "but," and it has made me fall from the symbolic world of language, myth, religion, and art to the "natural world" of physical reality, the immediate reality of man, the most characteristic and highest product of culture.

Anything that opposes the universal tendency of accretion to human knowledge (not simply personal, individual knowledge) is a deviation, an evolutional deviation and also a moral deviation, says Ladislav Kovac. To doubt organized science in the name of the "natural world" is to endanger the survival of the natural world.

The poet W. H. Auden understood this, as is clear from the definition of art he once gave me: Art is spiritual life made possible by science. I took it very personally.

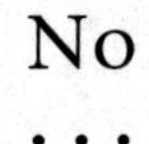
No
. . .

The Birth of Sisyphus

Blood under his feet,
overhead the beating
of mother's heart.

Naked, on the naked earth.

Blood will beget a boulder,
mother will turn into a mountain.

The gods will be mad as hell,
the creed will be gone with the wind.

The birth certificate will go missing,
the nail scissors lifted by a petty thief.

In the computer a virus will bloom like a hare's ear cabbage,
the power will be cut, Michael Jackson switched on.

Over the city, smog will sit heavy as wedding cake
and drunken shouts will stifle Solveig's song.

Global whining will be advised
and the continental plates will shift.

If an avalanche would come
the fuss would be over, but so far
it's just shift after shift,
stepping in shit,
love resembling death.

Guilty, on the guilty earth,
Sisyphus,

and what did you say his name was?

MIROSLAV HOLUB
translated by Rebekah Bloyd
with the author

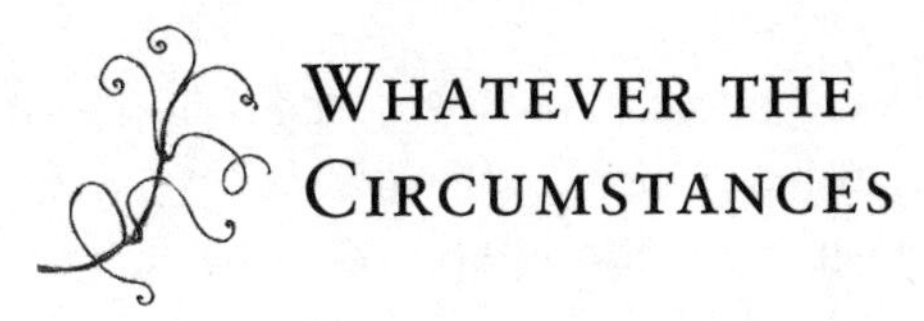

Whatever the Circumstances

The essential trait of an experiment is that it can be repeated whatever the circumstances, that is, under any conditions that are not integral components of the experimental system. If you inject allogeneic cells, you will get a cytotoxic response in May or in November and in Senegal or Stockholm; when you run a colony of nonobese diabetic (NOD) mice, you will get diabetes in 20 to 100 percent of the mice aged thirty weeks or more, without reference to the fact that your boss is a blockhead. Even in the less pictorial sciences, results occur without any fuzzy logic. Classical mechanics can predict and prove the position of planets, cannonballs, and baseballs, and quantum mechanics can predict the wavelengths of light emitted from the atoms in question without respect to any postmodernist assumptions: that is, whatever the circumstances.

In the human domain, however, it's indomitability, not repeatability, that counts against circumstances. In all times and places, there have been fairy tales, and their plots are generally about some magical solution to the problems of circumstances, as in the case of the poor girl who had to bring strawberries from the January forest. Michelangelo painted the dome of the Sistine Chapel despite the circumstances, and we may even reach Central Europe if we are ready to cope with all the circumstances.

Maybe that kind of indomitability requires some crazy stubbornness, or monomania. The results may be ambiguous, but they are the only kind of results we can expect in this world.

In the Amargosa desert there is a most striking example of indomitability: the Amargosa Opera, in Death Valley Junction. Maybe it isn't an opera, just an indomitability called opera, but there it is thirty miles from Shoshone Village, ninety miles from Baker and fifty miles from Beatty, Nevada. Death Valley Junction contains a couple of drab sheds that have survived from a borax-mining operation, and a sort of horseshoe-shaped one-story motel, inhabited by ghosts. Overhead, the burning blank blue sky or the depths of the black universe with exploding stars. There is no stop sign or traffic light at Death Valley Junction, and only inspired or knowledgeable drivers would bother to pause and look around. They would find at one end of the horseshoe a small sign announcing "Amargosa Opera—Performance Tonight." Tonight means Fridays, Saturdays, Sundays, and Mondays from October to May. At other times, obviously, it is so hot here that even the smallest trill, not to mention an aria, is precluded.

Except for the sign, there isn't a trace of life in daytime Death Valley Junction. I crawled through a window hole into one of the motel rooms, but it contained only some sparse rubble and a spider from the 1950s who was obviously afflicted with senile dementia.

So we returned in the evening. "We" included a very beautiful young lady. Beautiful young ladies are a tremendous motivation for exploring death valleys, operas, and all other circumstances. Near the Performance Tonight sign there were already a couple of motorcyclists, with their Jurassic helmets; the cars with daddies, mommies, and kids began to arrive. We kept assuring each other that there really would be an opera that night, and we kept watching something that resembled an entrance, in case something was going to happen and there might not be enough seats. Our faith was strengthened by some illuminated windows nearby. Just before eight o'clock an air-conditioned bus arrived from Furnace Creek. It was filled with tourists who, trapped in the middle of a dry dusty nature, seemed to be looking forward to a cultural experience. A red-faced man in a cowboy hat, obviously full of inspirational fluids, was noisily and enthusiastically praising the impending cultural experience, without much attempt to discriminate between an opera in the Amargosa desert and an amargosa in an opera desert.

Now lights appeared at the entrances and, like a deus ex machina, an elderly tall man in black velour appeared, sporting a huge red bow tie and the complexion of a music-hall comedian. He collected ten dollars from each of us. Upon payment, we found ourselves in a real theater auditorium whose lavishly decorated walls and ceilings, mostly gold and purple, illustrated scenes from Shakespeare's plays, with illusory balustrades and balconies, plus quotations that recalled their Latin origins very imperfectly.

We had hardly had our fill of looking at all these endearing frescoes, friezes, and fringes when the old-fashioned comedian appeared before the curtain and explained the origins of the Amargosa Opera as the inspiration of New York artist Marta Becket, who had discovered the abandoned building while passing through in 1967, had restored it, had decorated the entire theater herself (being a painter), and, since 1969, had been appearing here in ballets (being a ballet dancer) and in pantomimes (being also a mime).

We had hardly finished digesting all that when the ancient comedian was among us selling purple and gold programs for the 1984–85 season. Then it grew dark, the curtain opened, and before a Renaissance-rococo-Empire-style backdrop the prima donna herself, in a richly undulating sky-blue garment, nobly and gracefully fluttered and stamped to the accompaniment of a recorded Tchaikovsky romance. Following brief applause, the curtain opened again and she reappeared, this time to a calliope version of music by Juventino Rosas and wearing an abundantly feathered dress of a dawn-pink shade. We were all so deeply impressed by the lightning swiftness of her costume changes that the man in the cowboy hat was moved to shout "Bravo" and other enthusiastic vocalizations. In the next ballet, a dramatic vignette called "The Rumour," featuring the music of Alfredo Casella, the artist appeared successively as the First Woman, the Second Woman, and then the rumored Cabaret Dancer, each time in a different pea-green costume. Then came Sibelius, followed by intermission, during which the red-faced cowboy hat man noisily demanded enlightenment on the lightning costume changes, but without anyone's paying much attention to him.

After the intermission, the old-fashioned comedian introduced us to the charms of pantomime. The evening's program had a piece with the promising title "The Mortgage." In this, he told us, he would appear as a stage manager, seated in an armchair to one side, reading newspapers and commenting on the action. The artist appeared as a Dance-Hall Dancer, an Insurance-Man, a Tax-Collector, a Banker, an Attorney, three Sisters of the Original Dancer, a Villainous Mortgager pressing for payment, and a Ghost of the Dancer's Grandma, all these successively and often more than once. The action was tremendously varied, and not only were dresses changed, but beards and mustaches were stuck on; top hats and bowler hats, as well as frock coats, were donned and removed. The Villain's name was Thaddeus Tempest the Second, and the stage manager invited us to join the action by booing when we heard the sound of thunder (produced by himself), when we heard the name, or when Thaddeus appeared in person. The audience's booing grew progressively more enthusiastic, and the cowboy was shouting so vigorously by the end that he made the musical accompaniment (a quintet of French horns alternating with an organ) plainly superfluous.

Immense applause greeted the two actors as they took their bows. Later, all of us were invited to a reception in an adjacent Amargosa Opera exhibition hall, where the artist was offering her canvases, graphics, and postcards, all at rather considerable prices and provoking the "illusion du déjà vu" described by the French psychologist Janet, déjà vu in this case mainly of Edward Hopper paintings. Marta Becket had changed to everyday clothes, whereas the music-hall comedian elected to stay in his stage attire. They indicated that the two of them and a maintenance man made up the entire registered population of Death Valley Junction, which meant that Mrs. Becket was at the same time the mayor, police chief, and commander of the fire brigade, while her husband acted as the postmaster. As a result of keeping their town in good shape, they were expecting official state support. We had no doubt it would be forthcoming and reflected that in due course they might even establish a choir and an ice hockey team, given their unyielding spirit . . .

Then we drove to Beatty, through a deep and untethered desert night, with the silence of the not yet born and the lightness of the already dead around us.

We made no jokes about the opera. It was clear that indomitability and unyielding spirit are a lot easier to define than art is. And it isn't proper to make light of indomitability. If it's possible to dance to Sibelius in a ruin in Death Valley, it should be comparatively easy to have strawberries in January in our postcommunist forests, and to have philosophy and science even in institutions headed by blockheads. It should be possible to measure cytotoxic cells and wavelengths of emitted light even in postmodernist conditions governed by popular fuzzy logic.

Whatever the circumstances, I suggest, we should remember that we are, basically, in Death Valley Junction, and we should behave accordingly. The results may be ambiguous, but they are the only kind of results we can expect in this world.

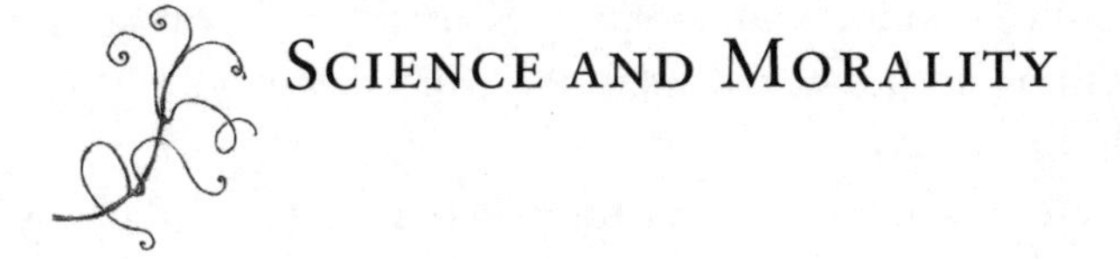

Science and Morality

It may not be obvious at first sight, but in modern societies of the so-called Western type, science not only plays an indispensable role in discovery, education, and organization; it also carries a moral force. This force is captured in Blaise Pascal's impressive notion that to think properly is a moral principle. Jacob Bronowski calls it "the habit of truth":

> The habit of testing and correcting the concept by its consequences in experience has been the spring within the movement of our civilization ever since. In science and in art, and in self-knowledge, we explore and move constantly by turning to the world of sense to ask, Is this so? This is the habit of truth, always minute yet always urgent, which for four hundred years has entered every action of ours.

It may not be obvious at first sight, but in the age of electronics, global communications, and computer solutions, it can be dangerous to think that suprapersonal, basic, and general problems can be dealt with on first sight, or that we can be led in our mental processes by this first sight or first intuition, or even that we can think and decide, as isolated individuals.

On the contrary, expert, collective decision making (collective not by setting up committees, but in the sense of making use of the best accessible technologies) is essential in a modern society of the Western type. And this essential activity cannot be denied and undone and changed by any wishful thinking, ritual, referendum, or dictate. "A civilization," says Bronowski, "cannot hold its activities apart, or put on science like a suit of clothes—a workday suit which is not good enough for Sundays."

Richard Feynman says very forcibly that "science is a long history of learning how not to fool ourselves." Not to fool ourselves, not to cheat ourselves, but to try to attain the objective (i.e., universally verifiable) truth, which is not a truth inside us but a truth among us—that is one of the few values of present-day life.

Science is an attainable truth—and therefore a moral value—even if we embrace another faith beside or above it (with the exception of the sometimes fanatical antiscientisms, promoted by unfurnished minds). Science rids us of demons sprouting in the fertile soil of fear of progress, which includes human liberation from natural fate, from oppression, from disease, and from inane minds. Science rids us also of the demons created by science itself. That is the most perfect definition of the notion of science and of its distinction from all the other cultural and civilized processes.

Kuhn's "second science," based on the simple observation of natural phenomena without a prejudiced mind and without the aid of devices transcending our senses—the "second science" as the function of common sense and its logic—leads to a mechanistic reductionism, which was the measure of its own limits. Hence arose Laplace's demon:

> The spirit who would know in a certain moment all the powers causing a stir in Nature and all the correlated positions of matter . . . genius enough to analyze this knowledge, he would find one formula including movements of the most voluminous bodies of the Universe as well as the movements

of the lightest atoms, for such a being nothing would be uncertain, he would see by his inner vision the past and the future at the same time.

Most laypeople accept Laplace's demon as guiding science in individual intuitive experience and the workings of artistic sensibility. In science, however, Laplace's demon was banished sixty years later by the thermodynamic laws and their statistical nature: neither in a geological and physical nor in a biological organization can the future states of systems be predicted.

What is left is the less graphic Maxwell's demon, capable in a mental experiment of bringing a set of molecules into the desired condition by an intelligent intervention, to change chaos into order and ineffective energy into a creative force. After this demon was expelled by information theory—that is, expelled from science—to function as a metaphor of poetic creativity in the broadest sense of the word, the scientific ground appeared. In its depth and in its entirety this ground seemed beyond the rational reach of the human individual, unless this individual was a professionally trained part of the planetary scientific context or unless, on the other hand, he did not view the world as a closed system, in the old-fashioned manner. In closed systems rigid causal relations rule; the system is influenceable from within, like an office agenda, which can be solved and eliminated piece by piece by dint of hard thinking. This last illusion was, among other things, the origin of Marx's demon, as demonstrated with uncommon elegance by Ladislav Kovac in 1995 in *Forum of Science*.

This last demon, however, met a particularly lamentable fate, because too soon after his birth he was thrown into the chilly Eastern world, where he froze and abandoned science to become a verbal instrument of power and even, reversing direction, a parasite on science, namely, on Soviet biology, medicine, physics, and general scientific theory—not to mention the liberal disciplines. The fact that scientific societies and idiosyncratic scientific structures finally got rid of him is one of the obvious proofs of the self-cleaning scientific process, which comprises inseparable moral aspects. A scientific society is based on respect and dissent: it is by definition a democracy.

In addition to truthful thinking, there are two other moral components of scientific practice.

The first component concerns the very method of getting knowledge. In *Science and Human Values* Bronowski writes:

> How do we get this knowledge? By behaving in a certain way: by adopting an ethic of science which makes knowledge possible. Therefore, the very activity of trying to refine and enhance knowledge—the discovering of "what is"—imposes on us a certain norm of conduct. The prime condition for its success is a scrupulous rectitude of behavior, based on a set of values like truth, trust, dignity, dissent and so on.
>
> Science confronts the work of one man with that of another and grafts each on each, and it cannot survive without justice and honor and respect between man and man. Only by these means can science pursue its steadfast object, to explore truth. If these values did not exist, then the society of scientists would have to invent them to make the practice of science possible. In societies where these values did not exist, science has had to create them.

This is nothing new: long ago Goethe, who had his share of specific knowledge, said that we need the categorical imperative in natural sciences just as we need it in ethics; but this may seem new or paradoxical to people who regard the natural sciences as the cynical lackeys of totalitarian systems or as the arrogant perpetrators of civilization's failures and disasters—ecological ones, for example.

This is, let us say, a misunderstanding: every totalitarian system, including the one that just ended, is antiscientific, especially when it loudly advertises science and proclaims its "scientific ideology," usually derived from no science at all, but from the appropriate demon. Science, any science, is located outside the horizon of such a system. Science, for a leader of the Eastern type, especially the well-indoctrinated one, who carries the parasitic ideological demon, is no more than what the leader himself understands, not the Spirit of the Earth. Thus he creates a puppet theater of the suitable and the unsuitable, admissible and inadmissible, "scientific" and unscientific. In the case of Marxism, this mix constituted an "Oblomov" attitude toward the evil bourgeois pseudoscience and managed not only to neutralize and annihilate philosophy and sociology, but also to absorb genetics, Virchow's pathology, psychosomatic medicine, chemical theories (resonance), and relativistic physics and cosmology. The ideologies of "suitable," acceptable science give rise to a lamentable shadow play pretending to link "progressive science" "more boldly" with practice, a shadow play in which progressive plans and figures are projected on academic screens, while behind them the scientists continue their independent existence, a hidden and scandalous existence to ideologists and technocrats because it is unpredictable. Under normal circumstances such independent conduct is called basic research.

A freelance ideologist, on the contrary, welcomes metaphysical and paranoid sciences, exorcism, parapsychology, and the natural cures of "alternative medicine," which have a single advantage: they are easy to understand, even for an ass. The secret police and espionage network of totalitarian systems of the Eastern type tried routinely to steal and exploit the results of "bourgeois pseudosciences" and technologies, while back at home they enthusiastically promoted charlatans promising everything from cancer cures to the detection of foreign submarines and flying saucers. Religious healers, who in America and nowadays even in my country operate the charitable

"Hallelujah" business, making in some cases millions of dollars, would be welcome guests of any communist regime if they omitted the good Lord and substituted particles of spiritual energy.

In view of this ideological background, anyone except cops and charlatans must realize that the ideas and laws of basic research have nothing to do with power, for a simple, fundamental reason: that an Eastern political leader—owing to his constitutional laziness—understands them no better than does a creation-science evangelist who has trouble with the American IRS because of his Sunday TV profits.

Bronowski's standard of conduct is built into basic research and can be disturbed by the authorities only in secondary matters—for instance, in the 1950s the obligation to quote at least one Soviet source in one's bibliography. It cannot be disturbed in its essence. One proof is the Czech participation in the discovery of immunological tolerance. In the early 1960s, Milan Hasek obtained the right results only because of the above-mentioned standards of laboratory work, although he and the whole institute were stifled under the glass bell of Michurinian biology; in the stale air of our biological institutes, ideological lightning keeps striking Monod, Medawar, Selye, and Wiener.

Interpersonal relationships in many scientific institutions can be quite petty at times because researchers, like other citizens, are part of the overall social situation, and the standards for laboratory conduct are—unfortunately—not transferable into the standards of competition for cushy jobs, easy money, and fat grants. Fortunately, the reverse is not true. Standards derived from the overall social situation cannot be transferred into specific thinking itself and into experimental questioning; in this respect, Bronowski's standards are constitutive features. An experiment is either an experiment or a fraud. Where standards of scientific conduct do not exist, science must create them, even going against the nature of the researchers themselves; they must realize, sooner or later, that nothing can be achieved without standards. That is why we speak scientifically about science and morality, not about scientists and morality.

If failures of civilization and the endangering of the planet were due to science, rather than to the political organization of life, we would expect a certain correspondence between the development and intensity of scientific research and the ecological status quo in a given country. In that case, after forty years of the "planned development of our scientific-research base," we Eastern Europeans should find ourselves in a green paradise undisturbed by emissions, beta radiators, or genetic engineering.

There is another factor that can be called moral, in the historical sense of that word. It is the fulfillment of a basic human need or obligation: inquisitiveness, curiosity, the need to know for sure, to perfect oneself, the need not to live in vain. The spirit needs to play, to make mistakes, to correct mistakes. He who admits at least once that an evolution of the human spirit exists, from the cutting of stones to thermodynamic laws, can understand Kierkegaard's words: "It is not true that a scientist goes after truth. Truth goes after him."

And this is related to the second principal moral component of scientific work, the public, civic value of practicing science and obtaining scientific results. In a nation that has adopted a Western-type culture and accepted the respective hierarchy of values, science contributes to national self-esteem. Science points out and influences general psychological possibilities, the level of national intelligence, the universal ability to organize, implement, and produce, to concentrate on a vital project, to think logically.

Victor Weisskopf commented on this:

> The value of fundamental research does not lie only in the ideas it produces. There is more to it. It affects the whole intellectual life of a nation by determining its way of thinking and the standards by which actions and intellectual productions are judged. If science is highly regarded and if the importance of being concerned with the most up-to-date problems of fundamental research is recognized, then a spiritual climate is created which influences other activities. An atmosphere of creativity is created which penetrates every cultural frontier. Applied sciences and technology are forced to adjust themselves to the highest intellectual standards which are developed in basic sciences. This influence works in many ways: Some fundamental students go into industry, where the techniques which were applied to meet the stringent requirements of fundamental research serve to create new technological methods. The style, the scale, and the level of scientific and technical work are determined in pure research; that is what attracts productive people and what brings scientists to those countries where science is at the highest level. Fundamental research sets the standards of modern scientific thought; it creates the intellectual climate in which our civilization flourishes. It pumps the lifeblood of idea and inventiveness not only into the technological laboratories and factories, but into every cultural activity of our time.

Although Weisskopf's words, written in 1965, reflect the optimism of their time and come from the perspective of a society in which science has reached the highest level, they can also refer to the situation in which rather little works, but in which there exist possibilities of an organic, natural, and positive feedback in a circular sequence: residual national talents and skills → science → technology → production and organization → social consciousness and morality → development of talent and skills.

Possibilities are always present for further development of the third science and of a new society, one where morality is not based on the thought processes of someone like a certain corporal in our army training course in the 1950s. That corporal believed that the values of sine alpha from plus one to minus one were a pedantic bourgeois anachronism, that in the army it could have a value of up to five, and during the Great Patriotic War and its powerful and moral awakening explosion of willpower, it had reached values of plus ten.

Aleksandr Herzen, a liberal thinker exiled from Russia in 1847, wrote that man and science are two concave mirrors that permanently mirror each other.

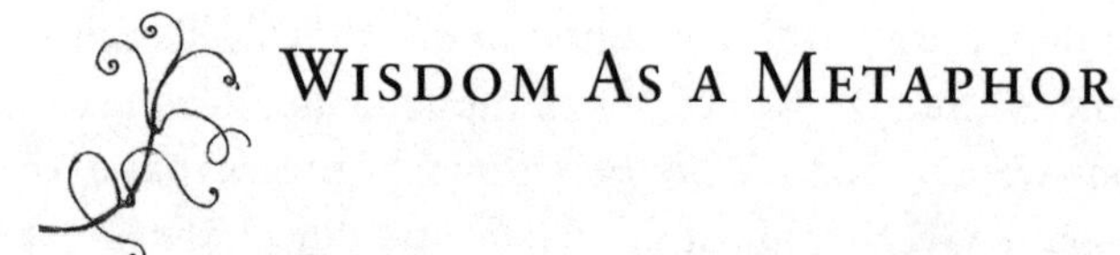

Wisdom As a Metaphor

AMONG THE POSITIVE HUMAN QUALITIES, wisdom is the most controversial. Is it in fact positive? Is it something that I could attain? That I'd like to attain? Everybody wants to be good (whatever that means). Everyone would like to be happy (on his or her own terms). But the idea of being wise, or sage, seems peripheral, something most people would deem desirable, but "not for me." I asked several people about this. None of them had a plan for becoming wise. They thought wisdom was something inveterate, or very old-fashioned. Most of them just wanted to be successful, period, with no further stipulations about being wise or foolish, clever or dull, reasonable or irrational.

Still, we do find people around us whom we consider more or less wise. We speak of unwise behavior, of wise or unwise solutions. We have the love of wisdom—philosophy—an agreeable and venerable remnant of stoicism. And we have a general sense of the meaning of the word *wisdom,* a quality we are apt to attribute to individuals other than ourselves.

Wisdom is a kind of virtual reality, and virtual reality, as such, is a metaphor.

Wisdom is a philosopher's tool, or toy.

In 1920 Karel Capek wrote a story called "Agathon; or, On Wisdom." In it, Agathon comes to Boeotia to deliver a philosophical lecture. He is disappointed that there are only a few listeners assembled, and he comments: "I am aware that you Boeotians are preoccupied with the local elections and that this is no time for wisdom, not even for reason; elections are an occasion for cunning." He then goes on to analyze the three words *cleverness, reason,* and *wisdom.* All three denote some intellectual capacity, but wisdom is distinct from cleverness and reason because there can be thoughts that are neither smart nor reasonable, but that are nonetheless wise. "Cleverness," says Agathon,

> is as a rule cruel, malevolent and selfish; it looks for weaknesses in a neighbor and uses them for its own profit. It leads to success.
>
> Reason may be cruel to the man, but it is just to the goals; it looks for common gains; if it finds a neighbor's fault, it attempts to correct it. It leads to improvement.
>
> Wisdom cannot be cruel, since it is kindness and sympathy in and of itself; it does not seek common gains, loving people too much to love a goal beyond them. If it finds a weakness, it forgives and pities; it leads to harmony.
>
> Yes, wisdom is sorrow, in a way. Reason can be wisdom. Reason is in the deed, wisdom is in the personal experience.
>
> But wise poets and artists may be able to put that experience in their works. . . . This is, then, the special value of art, second to nothing in the world.

With this, Capek's Agathon concludes his Boeotian lecture: since wisdom lies in personal experience, he has in fact nothing more to lecture about.

So Karel Capek's wisdom is not a capacity of mind, but a state of mind, a little sad, somewhat helpless, but harmonious and communicable only by wise poets. I am not sure whether Capek would have proposed the same definition eighteen years later, at the end of his life, or if he had lived even longer, after seven years of Nazi occupation and a subsequent forty years of overwhelming Russian brotherly help. In other words, I'm not sure that wisdom can be independent of place and time. I'm also not sure about the biological phenomenon of the wise poet, with eternal and universal value.

Years ago, without any sage intentions, I wrote a poem called "Wisdom":

> But poetry should never be a thicket,
> no matter how delightful, where
> the frightened fawn of sense can hide.
>
> And this is a story of wisdom,
> allied with the roots of life
> And therefore
> in the dark
> and blind.
> A small boy not yet bound
> by the hempen fetter of speech,
> With only ten jingling words
> on his tongue.
> But already in the iron shirt of sickness,
> heavier than a man could bear upright.
> In a white box resembling a glass mountain,
> from which all knights
> tumble head over heels—
> There is nothing in the mind that
> hasn't been in life.
>
> (At that time
> tuberculous meningitis
> still occurred.)
> On Christmas Eve he got his first
> toys, a giraffe and a red car.
> And in the corridor—far from this
> continent—stood a Christmas tree

with tears in its eyes.
There's nothing in life that
hasn't been in the mind.

And the little boy played, amidst
symptoms and in the blue valley
of the fever chart,
And between two lumbar punctures,
not unlike
being crucified,
He played with his giraffe and his red car
which represented
his crown jewels of time,
all Christmases and
all the Punches of the world.
And when we asked him
what else he would like
He said with a feverish gaze
from the beyond:
Nothing else.
Wisdom is not in the multitude
but in the one.

(At that time
basilar meningitis was still
fatal.)
It was a very white Christmas,
snow down to the roots,
frost up to the sky,
And the glass mountain's tremor
perceptible under our hands.

And he just played.

Even now, many years after having written that poem, I still think it is about wisdom. That means that wisdom is a restraint, a moderation and a silence, not necessarily obtained by meditation, reasoning, or reading the classics. And that one kind of wisdom, at least, is connected with, and related to, the very roots of life.

But unlike Capek's philosopher, the fatally ill boy in the poem can't do anything more. Healthy philosophers can and should use their minds for more than internal peace and poetry, since they have a choice other people don't have. My poem is about the limiting situation in which wisdom is the last resource, the act of a simple mind that has no choice. That kind of wisdom is a function of the existential situation. It's a heroic wisdom that makes us think of Brecht's Galileo and his wish that we had no need of heroes.

This kind of wisdom is a state of mind, perhaps the ultimate state of mind, and it lacks attributes of sageness and learning. It's not the highest perfection of man, as in Confucianism; it's simply a kind of mental endorphin: a clearly metaphoric (but also obvious) application of the word.

In any case, we have now looked at one kind of wisdom. There are obviously many other kinds. I was recently struck by an article by our philosophizing biologist Stanislav Komárek, who said:

> The desire to achieve wisdom is very rare and unconventional in our era, so that . . . a short definition of what it is like is needed. The best definition would be that wisdom is the capacity of discerning what is essential in the given situation and what is not. . . . A wise man can use the essential for some benefit. . . . Modern science was, in its oldest history, related to the quest for wisdom, but now it has become fully emancipated from it. In the huge, self-centered system represented also by the system of institutions and research agencies, the question about the essential is perceived as absurd. . . . On the contrary, the machinery of scientific institutions represents a good protection against this question.

I am afraid this is a typical misuse of the wisdom metaphor. On the one hand, discerning what is essential in a concrete situation is a basic biological capacity. Every frog must discern the essential marks of the prey and food, of the predator and angel of death. Each cell must discern essential and nonessential messages by exposing the pertinent receptors. I once labeled the immune system "the inner wisdom of the body" strictly in the metaphorical sense of the word. Wisdom and immune response can be said to share a characteristic and a value.

Here, then, is a second metaphoric use of the term *wisdom*.

On the other hand, if organized modern science has to be excluded from the realm of wisdom, one must ask whether the only wisdom is that which concerns human spiritual and metaphysical aspirations, or whether it can concern all the problems of human existence, such as survival, technical progress, and health. I can't accept the idea that wisdom is all in the workings of the mind, that it's something nonoperational, impractical, or even esoteric.

Moreover, the exclusion of science is due to a basic misunderstanding about its practice. I can't imagine any modern scientific discipline functioning successfully without the accompanying question of what is essential to each concrete, state-of-the-art situation. Science is the art of the soluble; it requires not only education and experience, but also soundness of judgment, restraint, and moderation. These attributes are needed to define the soluble and to escape such phantasmagorias as Russian Marxism-Leninism and "creationist science." In this sense, wisdom is one of the presuppositions of modern science. We locate it in modern and postmodern society, but Belinsky, the Russian literary theorist, had already noted, some hundred and fifty years ago, that "to scorn and degrade science, reason and art is unwise."

The exclusion of modern science from the realm of wisdom is probably also due to the routine misunderstanding that takes contemporary science as a discourse, made up of words, a mode of philosophical interpretation. Contemporary science is a wise technical instruction, wise *because* it's technical, in how to use human capacities on this human planet. It is in fact the only component of culture that cannot be obscured by words and by vague notions about what may be generally essential. Contemporary science and technology are the means of human survival.

Definitions of wisdom that exclude science, at this stage in the planet's history, are just lazy, persistent metaphors.

In addition, the sense of what's essential changes century by century, year by year. Right now, science is the only realm of human intellectual activity in which we have something real that we can be clever and wise about, to paraphrase Peter Medawar. And human intellectual activity, even when it's called wisdom, is not a kind of folklore. It is the core of human existence in this more or less comprehensible world bulging with hard facts and soft, obscure words.

Maybe we can clarify this if we turn for a moment to the roots and traditions of the notion of wisdom. According to *Webster's Third New International Dictionary,* wisdom is the effectual mediating principle, or personification, of God's will in the creation of the world. The expression "wisdom" was applied metaphorically to the teachings of ancient prophets in Babylon, Egypt, and biblical Palestine, teachings that dealt with the art of living and, sometimes, with philosophical problems. Wisdom in this sense constituted a class of literature, as in the Old Testament books of Job, Proverbs, Ecclesiastes, and the Song of Solomon.

The idea of divine inspiration has remained associated with the wisdom metaphor in popular understanding ever since. Hence the exclusion of "wisdom" from the personal aspirations of ordinary people, with some notable exceptions that take the form of paranoia.

There are other lexical meanings of wisdom. There's the sense of its embodiment in a literary form, such as aphorism or personification. More recently, it is said to suggest accumulated information, learning and its intelligent application, insight and sagacity, good sense, sound judgment, prudence, and sanity. It's obvious that the exclusion of science from wisdom and wisdom from science is an error that is already built in at the lexical level.

But the accumulated and persistent lore of wisdom metaphors has caused its share of misunderstandings at the level of culture and politics. Culture and its carriers, the intellectuals, may attain and bear some wisdom, while politicians and technologists may bear some power, but almost never wisdom. Power means corruption, intellectuality means deep insights and sagacity. This simplistic polarization is most clearly visible in the new democracies of Central Europe: it happened here before our eyes, as in time-lapse cinematography of cloud formations or cell movements.

Within just a few years, people with almost identical biographies and curricula differentiated themselves into pragmatic politicians or independent, free, and responsible torchbearers. The latter were always in opposition, always idealistic (in Péguy's sense of the word, that is, idealists would have clean hands, if they had any). These torchbearers kept clear of the political machine and its cogwheels, maintaining their wise independence.

Of course, wise independence is relative. One may be independent of the administration, of the political parties, of the prime minister, and of traffic regulations, but this in turn implies that we'll be more dependent on our own *idola mentis,* idols of the mind, on the cultural *idola fori,* idols of the marketplace, and on our own personal instincts and neurohumoral deviations. I'm not sure of the extent to which personal instincts and deviations can be said to constitute wisdom. In many cases they lead to trespassing on the obvious limits of freedom, into anarchy.

The real self of a free intellectual is based on integrity. Integrity is in fairly short supply, actually, among both the new politicians and the wise intellectuals. It's just that pretending integrity and wisdom is a lot easier for intellectuals who, as a rule, escape the scrutiny of public opinion and insensitive journalists. It's also true, as the Slovakian advocate of cognitive biology Ladislav Kovac has pointed out, that intellectuals have appetites that they frequently can't satisfy. Compensating for them develops moral complexes and anxieties that interfere with intelligence. We are not selected, Kovac says, to recognize the deep motives of our behavior, and intellectuals are among the most colorful examples.

The intellectual torchbearer opposes to the politician's illusion of power a homemade illusion of transcendence, for the pleasures of transcendence "we have our philosophical persons, to make modern and familiar things supernatural and causeless," as Shakespeare tells us in *All's Well That Ends Well.* "But the dream about the infinity of the human soul," writes Milan Kundera, "loses its charm in the moment when history takes the man in its grip." I don't know any example of a Diogenes who escaped history. With the exception of foolproof psychopaths.

In the words of the Czech philosopher Vaclav Belohradsky, there is a permanent confusion between the dichotomy *true-false* and the dichotomy *sacred-profane.* In his view, the concept of intellectual engagement as a fight for the sacred true world, as opposed to the false profane world, is the basis of the unwise intellectual opposition that may have led, in some historical conditions, to holocausts and gulags.

I assume that the wisdom metaphor applies to all professions. There is nothing new about it. Cicero had to point out some time ago that nine-tenths of wisdom means being wise at the proper time, and that nobody is really wise if he is not audacious. The Roman *sapientia* implied political deliberation and constituted the highest virtue only in combination with *virtus,* bravery. A little later, Brecht said that the only mark of wisdom in a wise man is his behavior.

Constituting a healthy Western intellectual requires not only a wisdom relative to rational discourse, knowledge, and introspection; there must also be operational thinking and an experimental approach, wherever possible, along with some administrative skill. Not just the wisdom of the long beard, occasionally prophetic, but also the wisdom of the firm hand.

This is not an academic postulate. In our recent history we have the example of the president and philosopher or social scientist T. G. Masaryk. The present president of the Czech Republic, the writer Václav Havel, meets Brecht's requirement well.

What we witness, however, among many intellectuals in the new democracies, instead of understanding and political activity, is a mass movement toward the Oriental metaphors of wisdom as passivity. Paradoxically, these come from the West and are corroborated by Western "experience"; for Czechs, Hungarians, and Poles, what comes from the West must be superior to anything else, especially if it was forbidden by the Reds. To get Zen, yoga, spiritual insights, and meditation all the way from India or China may be as suspect as getting Pavlovian medicine from Russia, but to get the same things from California is animating and fashionable. To join a sect from Utah is even more interesting than to join a sect from Uttar Pradesh. And so we have a tide of more or less commercial Indian and Christian Science gurus as splendid examples of a superior wisdom, offering easy medieval solutions to the problems of the twentieth century. A bearded, dark-eyed Indian prophet is held to be expert in everything from thermoregulation and oxygen consumption to eternity.

In this mode, wisdom is not only a metaphor, but also a paranoid message. Paranoid not from its place of origin but because of its destination. Paranoid as in the transplant of a surplus kidney to a patient with normally functioning kidneys. Nevertheless, Eastern wisdom, as a universal message from the universe, is supposed to be the deepest wisdom attainable.

I'm afraid that many poets subscribe to this mode of wisdom, turning Capek's good philosopher into an idiot. My child with basilar meningitis is protected, but only because he's dying.

Having come to this issue, we're at a continental divide. Is it possible to approach truth by meditation and introspection, or do we need all the information that is available to an intelligent citizen, one who did not, and does not, want to forget what he learned in high school? For a Hindu guru in a remote Indian village, truth and wisdom may be acquired by looking up to a blue sky over a stupa or a ruin. For a Hindu guru in Europe, and for any European sage, anywhere in the world, scientific and technological information is the inevitable first step toward understanding the world and surviving in a dynamic civilization. Replacing information with meditation is as silly as replacing a jet engine with candles. In our culture, the velvety Eastern wisdoms are misplaced wisdoms, although they may be useful in curing some mental disorders.

This misplaced wisdom has very tangible consequences. Supernatural powers guide everyday life. Psychics are asked to help police in discovering terrorists. Faith healing becomes a rewarding profession; in Czechoslovakia we have 20,000 registered healers (and many more unregistered), as compared with 7,500 in the United Kingdom. It's a lot cheaper than official medicine, the argument runs, and Philippine psychic surgery is shown on television with the shrugging accompanying comment, Who knows? Geopathogenic zones are abundant, and so are commercial devices for blocking them. Even the top politicians have this weakness for all kinds of alternative science. Oriental doctrines must be superior: look how old they are.

The "green" mythology is based on the notion of the everlasting wisdom of nature, again with an Oriental touch, and stemming from the postmodern longing for the relativization of science and the condemnation of all new technologies (always excepting the technology for transmitting the pertinent propaganda). The wisdom of nature as part of the green mythology is the most far-fetched of all the wisdom metaphors.

One trouble with the wisdom of nature is that nobody can draw the dividing line between nature and man: Is my neuron unnatural while the neuron of an opossum is natural? Is the neocortex of our brain unnatural? Is a housefly part of nature? Is a free-living spirochete different from a *Treponema pallidum* in Socrates' blood and from spirochetes that, according to Lynn Margulis, turned into constituents of our sensory organs, if not brain cells? Were Kwakiutl Indians (with their cannibal spirits) less unnatural than the white settlers who founded Portland, Oregon, in 1844?

Is an adenocarcinoma natural, and is an adenocarcinoma of the lung brought on by smoking less natural? Why do we kill *Corynebacterium diphtheriae* when it is so natural and in spite of the fact that it doesn't want to kill us, kills only because of the fact that it itself has caught a virus, a phage?

By *nature* we usually mean the landscape, which was changed by human activities hundreds of years ago. By *man* we usually mean the result of a natural anthropogenesis that created civilization instead of succumbing to natural disasters.

And yet, disasters are a constitutive component of that natural wisdom that is in turn an important force of planetary evolution.

We are not supposed to tinker with nature by genetic engineering of tomatoes and cattle and hemophilia. On the other hand, we *are* supposed to tinker with nature by fighting retroviruses such as HIV. We are supposed to observe animal rights after 11,000 years of domestication even though we are faced with these hard facts: that the only alternative to animal experimentation is experimentation on patients, and that 6,228 million people in the year 2000 must somehow be fed by scientifically improved livestock and crops, since the world grainland area per person keeps dropping—from twenty-two acres in 1950 to seventeen acres in 1995. The overpopulation of the place cannot be solved by letting some human populations die out, and certainly not by letting nature take its course, as Roman Catholic "wisdom" recommends.

In addition, we are supposed to admire the divinely wise design of our bodies, in spite of the fact that the heart is not the best kind of pump and the lungs are faulty by design (by a complicated phylogeny). And despite the fact that there is a minor but regular mix-up of endocrine cells in the mediastinal area, and that the descent of the testes is a comic developmental incident that makes mammals inferior in design to any annelid worm. And despite the fact that our wise immune systems make terrible errors every day.

Life is an offensive directed against the repetitious mechanisms of the universe, said Alfred North Whitehead. But this does not mean that it is the best offensive one can think of. What we mean by wisdom and reason is by definition superior to what nature meant. The metaphor is usually wiser than its author, said Georg Lichtenberg.

Listening to the green and pink fundamentalists, I always remember François Jacob's criticism of the wisdom of nature in the human case: "The development of a dominant neocortex supported by old-fashioned nervous and hormonal systems, in part remaining autonomous, in part placed under the neocortex's tutelage, all this evolutionary process looks very much of a patched-up job. It's a bit like fitting a jet engine onto an old horse-cart. Hardly surprising if accidents occur." In satiric terms, the brain is an instrument by means of which we think we think, said Julian Tuwim.

At the critical points, the patched-up job of natural evolution can be corrected by newly acquired human capacities based on information and erudition, constituting something like a limiting wisdom, as I suggest in the poem "Of Course":

Of course,
the first philosophy
is the philosophy of the liver, the kidneys,
heart muscle,
pancreatic islands,
red bone marrow
and stem cells,
infinite in their own fashion.

In the Socratic transplant program,
that discourse of body, knife, and electronics,
one spiritless Self is crossed with another,
while an automatic virtuoso
plays solo violin, accompanied by an orchestra
with muted trumpets.

Mozart ought to have gotten a kidney,
Spinoza was waiting for new lungs,
and Kierkegaard needed a heart,
or at least a valve.
All in vain.

Because
that bloody flesh

in the claws of the cave birds of narcosis
is the only real wisdom–
new, real,
and transmittable.

We are in a limiting and ultimate situation, where wisdom is revealed not in words but in action. Only informed and trained–and therefore collective–wisdom can save lives and be real wisdom.

I have mentioned five metaphoric wisdoms: wisdom as a state of mind; wisdom as an endorphin, as resignation; wisdom as a basic biological function; wisdom as a traditional function of prophets; and wisdom as the mythic green wisdom of nature.

Even poets should be aware of what may be wisdom, what may be metaphoric wisdom, and what the rules of the game are. Here I cite my prose poem "Zito the Magician":

> To amuse his Royal Majesty, he will change water into wine. Frogs into footmen. Beetles into bailiffs. And make a Minister out of a rat. He bows, and daisies grow from his fingertips. And a talking bird sits on his shoulder.
>
> There.
>
> Think up something else, demands his Royal Majesty. Think up a black star. So he thinks up a black star. Think up dry water. So he thinks up dry water. Think of a river bound by straw-bands. So he does.
>
> There.
>
> Then along comes a student and asks: Think up sine alpha greater than one.
>
> And Zito grows pale and sad: Terribly sorry. Sine is between plus one and minus one. Nothing you can do about that. And

> he leaves the great royal empire, quietly weaves his way
> through the throng of courtiers, to his home
> in a nutshell.

Without some relationship to the real world, poetry may turn into a verbal game and a babble about the more or less interesting self.

"Yet there are times," said Seamus Heaney in his 1995 Nobel lecture, "when a deeper need enters, when we want the poem to be not only pleasurably right, but compellingly wise, not only a surprising variation played upon the world, but a retuning of the world itself . . . (this is a need for poetry) . . . as an order 'true to the impact of external reality and . . . sensitive to the inner laws of the poet's being.'" It is compellingly meaningful that the Nobel Prize goes to a poet who sees poetry as an order and a function of wisdom in the literal sense of the word. In a time of postmodernist nihilism and a prevailing relativism in our poetry, irrelevant by virtue of its irrelevance, Heaney comments: "What I was longing for was not quite stability but an active escape from the quicksand of relativism, a way of crediting poetry without anxiety or apology."

It may be this aspect of poetry that Richard Rorty had in mind when he suggested that in the future, poetry may have more importance than philosophy. Richard Kearney, the Irish philosopher/critic, explains:

> The free play of imagination is unrivaled not only for poetics, but also for ethics. . . . When ethics is left alone . . . it is likely to degenerate into uninspiring moralizing. Ethics needs poetics as a reminder that responsibility toward the other person includes the possibility of the game, of freedom and joy, just as poetics needs the experience of the other person . . . and aims at this experience. . . . [Both are] an hermeneutic act of being-for-somebody-else.

It is, again, compellingly meaningful that Heaney and Kearney speak of joy and pleasurable rightness. Joy and pleasure are not states of mind well known in our latitudes, although they may be closer to actual wisdom than Capek's "kind of sadness."

The reason why some of us are sensitive to misplaced wisdoms and other metaphors, as well as to the practical aspects of wisdom, may be found in the forty years we have spent under the rule of arrogant and pompous ignorance. Josef Capek observed that the true opposite of wisdom is not stupidity but madness: pompous foolishness.

What is the sine alpha for intellectual communities with their idols of the mind and their idols of the marketplace?

There are moments when one appreciates wisdom as a restraint, a moderation, and a last resource.

A wise intellectual, wrote Albert Camus, is someone whose mind watches itself.

A Concert in Morelia

It could have been any place and it could have been any time. But it was in Morelia, in the state of Michoacán, in southwestern Mexico, and it reminded me of the way things are, any place and any time.

It was a meeting of the Grupo de los Cien, the Group of One Hundred, which means it was almost fifty of us, from all corners of the world, mainly the American world but some rare specimens imported from Europe.

The meeting was about ecology, which means that about five of us actually specialized in ecology and even understood what it is, while the rest of us emitted strong ecological sentiments, as is proper for artists. After some productive and fulfilling meetings, we were duly transported to the center of Morelia, for purposes of cultural monument visitation.

By this means we were transformed into a group of glamorous and intrepid travelers, distinguished by intense cultural curiosity, sporting a modestly glowing international halo and an acute sensitivity to Michoacán folklore combined with an Anabaptist hunger for knowledge.

Radiating our effulgence and signing the requisite guest books, we visited a string of churches, sacristies, statues, bumpy roads, museums, and aqueducts, arriving eventually in an old monastery that has been converted into a cultural center. Welcomed through loudspeakers—it was almost like the Lord talking to Abraham from on high—we were ushered through the ground-floor exhibition of Michoacán dance masks. These are the products of particular tribes who indulge in many colorful dances in the absence of sufficient numbers of taverns and television sets.

We were next taken upstairs to the gallery of contemporary city art. The gallery contained recent masterworks by local artists and artisans who had been deeply affected by the effects of European cultural imperialism. Most numerous were victims of Van Gogh, but there were also many who had been wounded by Monet, infected by Picabia, colonized by Picasso, and struck by Tapies. Not a single plain and honest Indian seemed to be left. Many of us, enlightened internationalists, made a rapid escape. Only some culturally decimated types remained and wandered around the monastery, having lost our guides.

Thus it happened that I arrived in a huge, roomy hall at the end of one of the vaulted corridors. A hundred empty chairs were waiting artfully arranged in rows, and at the far end, below windows, was a platform containing an orchestra. It was a children's orchestra, partly dark, partly white, solemnly clad, a hair bow here, a bow tie there, the conductor/preceptor in front. The violinists had their bows on their strings, the clarinetists, flutists, and other whistlers had their lips on their mouthpieces, and a little boy in the percussion section had a hand with a hammer raised. The conductor had his baton raised and was gazing back toward the entrance in a rather strange way, under his shoulder. The entire orchestra had their shiny black eyes turned in the same direction, motionless, like an orchestral still life. It was as if Sleeping Beauty had just pricked herself.

No international kiss was needed to set the scene in motion, though. Immediately after I appeared at the entrance, the astronomical clock was set in motion.

They fiddled and played all out, while now and then a black eye glanced up from the score toward the door, where the remaining ninety-nine intrepid cultural world citizens were supposed to emerge.

Four emerged, two lured by the sound of the music and two who did not care about the music but were searching desperately for a rest room.

Gershwin was rendered, and, without any noticeable pause, Johann Strauss, Ravel, Bizet, and Mussorgsky followed in a hectic sequence, driven by fear of the disappearance of the token audience. Black eyes followed us from above the instruments as we took seats with the utmost politeness and so as to cover as much of the void as possible—with the exception of one of the ladies in need of the WC, who discreetly but decisively retreated through the central aisle.

If improper cultural domination had reigned in the art gallery, a kind of revenge was enacted here: all the Euro-American composers were thoroughly defeated. Not much remained of them beyond occasional suggestions of theme and melody, along with disconnected rhythms and blurred scales.

I began to expect that after Mussorgsky they would intonate a Saltikov-Schedrin and execute a double Rittberger jump.

But I stopped picking up the sound track and became fully engrossed in the theater of the world. These children of Morelia had obviously been preparing for this moment for months. And these children of Morelia would remember this production, these thirty minutes of their lives, for twenty years to come, be it with tequila in hand or with little son on knee. This was the heroic moment for these little Indian souls and for the hypertrophic soul of their white preceptor.

Most of them would never again have a listener from abroad, not an international nor an intercontinental, probably not even one from Mexico City. Whether or not we were here by mistake, by chance, or by predestination, whether or not we kept our ears to the ground or were simply looking for the WC, we were here for the great moment of having the world listen to the children's orchestra.

There is nothing more important than listening, whether or not the music is worth a tinker's damn.

We were in that absurd empty hall, with the orange Mexican sunset streaming through the windows, not for the Group of One Hundred but for all the groups of people who can't manage to do much about the poverty and hunger in the Second and a Half or Third World, but who can still manage to listen, really listen, thereby inducing ten or twenty minutes of moderate happiness in the girls with bows in their black hair who won't make it to Carnegie Hall in a hundred years even if their fathers don't sell the fiddles to a pawnbroker within the next two years, as well as ten minutes of pride in the little boys who will fade back into the cornfields because the United States is a long way off.

So I was sitting there like a most interested and involved boulder, trying to neutralize, by means of deep enthusiasm, the essential emptiness of the hall.

And it occurred to me, in the middle of *L'Arlésienne,* that it was not an isolated or pitiful situation; it was a general and representative cultural situation. Is not the essence of all art that it is never fully accomplished? Convincing? Never perfect? So it's a frequent occurrence that we're interested boulders, impregnated by goodwill. We jubilate, even if small misgivings, miscues, misdirections, and mismatches are obvious, indeed, even intrusive. Even if the performance is a sort of well-meant forgery. Because it is performed by good, decent people, so that we applaud the decent, well-meaning people rather than the art.

So I was doing what I do on most occasions, only here it was somewhat more important. In *Broca's Brain,* Carl Sagan is very persuasive in arguing that religious feelings are based on prenatal and perinatal experience that is real and certain, but also blurred and vague, resisting accurate recollection. I would suggest in addition that cultural feelings and behaviors have something to do with the situation of the helpless infant, seeking contact with the nearest human being, identifying mother's heartbeat, alternating personal modes according to the stimuli provided by others, imprinted by the human patterns in the tiny perceivable universe called home. I feel that culture has more to do with human proximity than with other values, and much more to do with children, their games, and their troubles than with the bearded oldsters.

So that our situation in the empty hall was something like a cultural archetype. The archetype lasted half an hour. Then we were discovered by our guides, given away by that lady who had finally found the toilet. She and the rest of the internationalists suggested that such a concert constituted an encroachment on our freedom. And what is better for an artist than freedom, spiritual roaming, and dinner?

I did not think so. I thought that freedom without essential respect and decency leads nowhere. But we had to go, because of the democratic principle of majority rule and the fact of only one waiting bus.

We rose gingerly and crept along the wall toward the door; the sun had already turned purple and the orchestra was playing something that dimly approximated Leonard Bernstein. *West Side Story,* completely out of tune and scale, blew through the dark corridors of the ancient monastery. The kids played on, the preceptor went on conducting, perhaps only because the kids went on playing, even when nobody remained in the roomy auditorium, not even a mouse.

It sounded in the walkways, on the day stairs and night stairs and in the refectory and among the Indian costumes in the wardrobes, and it sounded relatively victorious, but I was not sure. There are so many victorious noises around, these days.

Let's say then that it sounded like a moderate defeat for all of us, including Leonard Bernstein.

A Scientific Horror Story

Reading through old scientific journals can be interesting. Browsing through the 1979 issues of an international medical journal, I noticed that one author contributed four separate papers. The first came to the editors on November 21, 1978, and was a sophisticated review article on inactive, dormant tumors. The second, from November 24, was a clinical study of the prognostic value of immunoglobulin levels in patients with Hodgkin's disease. The assessments were carried out on thirty-one patients and forty healthy subjects. Then, immediately after, on December 12, the very same author sent another study on finding cancer and embryonic antigens in patients with malignant and benign tumors. There were 289 of the first type and 261 of the other, with a supplementary control of 250 healthy subjects. And finally, on that same December 12, the very same author added another piece of work on serum immunoglobulins in acute myeloid leukemia; the work comprised findings in 107 patients. Thus, in just that one year, and judging from one journal only, at least 978 patients and a series of masterful, successful laboratory techniques had passed through this author's hands, along with 113 references to relevant literature and research.

This prodigious author, however, was exceptionally humble, citing himself only once and describing the medical facilities where he examined his patients very vaguely. In the November papers he cited his workplace as the Royal Scientific Society in Amman, Jordan, while in the December works it was the Reference Unit for Specific Proteins in Baghdad, Iraq. All this while his home address was 7511 Teal Run, Houston, Texas. In November he thanked the crown prince for financial support; in December it was the president of an unnamed institution. Yet the prince and the president had the same name.

It should be added that the level of all four works exceeded the average quality of papers in the given journal and that their English was also substantially better. The referees through whose hands the works must have gone prior to acceptance for publication were presumably satisfied, perhaps even delighted. Referees check neither the clinics nor the sources of financial support nor the authors' addresses. Nor could they, because individual works are sent to various referees. Then, to the extent that they knew the pertinent literature, they also must have been impressed by the fact that this author had published more than sixty similar, solid works over two years, suggesting a copiousness bordering on the miraculous. Overall, he had published in about twenty European, American, and Japanese journals.

And he would still be publishing today had a certain graduate student in Kansas City not discovered that work of his own had appeared under the name of Elias A. K. Alsabti prior to its appearance in the journal where he had sent it. E. A. K. Alsabti must have copied it before it was sent off—which was otherwise not his style. Changing your habits and style may be detrimental to your career.

E. A. K. Alsabti had hit upon a brilliant idea. He didn't toil away at the minute doctoring of tables and the fabrication of data. He didn't even toil at a single bit of clinical or laboratory work. Furthermore, he had no qualification or education suitable to such work. He made up his own university diplomas. On the other hand, he really did own a yellow, and later a white, Cadillac, and stationery embossed with a royal crown. He changed his workplaces in America with great foresight and skill. He had no problem with acceptances in the journals; where necessary, he covered the expenses himself, and his scientific résumé was a delectably growing list of published papers. In each new workplace he would simply study a number of journals, select published works, change their titles so that the computers of reference centers could not identify them, recast the data on the origins of patients, and, finally, exchange the original author's name for his own. He would then send these scientific papers to journals where he was sure they would not end up in the hands of the original author. It was not too difficult: eight thousand medical journals were available to him. Articles stolen from journals of the highest quality thrived, imperceptibly, in minor journals.

When he was found out, he had just been employed at the University of Virginia, and he didn't give up right away. He sent menacing letters to the authors of the revealing articles, threatening them with lawsuits and insisting that some unknown slanderer had put his name on the articles that unmasked him. Finally, when he had been exposed worldwide, the subject of editorials in the world's most prestigious journals and one of the protagonists of the book *Betrayers of the Truth,* he vanished.

One may suspect that he is appearing elsewhere under a different name. As the embodiment of one kind of literary license, Alsabti won't be forgotten soon.

From the medical point of view, it can be said that some 978 patients from our journal, as well as the thousands and thousands of others in Alsabti's reports, can congratulate themselves on not having actually passed through his hands.

From the scientific point of view, let's call it a horror story.

Not because it's possible to have a career this way, no matter how brief, but because only the quantitative dimension of his activity led to his exposure and demise. Had he satisfied himself with ten copied articles a year, the probability of discovery would have been significantly less. It was precisely some hectic trait that led to the crisis and catharsis of this tragicomedy. His case deepens my conviction that the process of scientific publication hasn't yet reached scientific exactness. Manifest loopholes still exist in the assessment of scientific performance.

It doesn't shake my conviction that a self-cleansing capacity exists in science, as a unique phenomenon in human enterprise. Science as a collective activity is highly moral, even though it may sometimes be the result of activities by individuals who are morally not very well developed. The results of individual effort, sometimes also called the creative process, quickly, perhaps even at the moment of their publication, become separate from the person. They can be supported by his or her renown, they can be jeopardized by his or her lack of it, but in principle they enter the impersonal, cybernetic, global "scientific consciousness" in which they take up a central or a peripheral position, milling around with and rubbing up against vast numbers of similar results. In the nexus of agreement a minuscule truth appears; in the nexus of disagreement and conflict, a particle of doubt. The chaff of deformations, fabrications, and deceptions slowly emerges. The consciousness remains more or less clean, even though nourished by the ballast of minds not entirely clean, minds that can damage the public reputation of science but not science itself.

From the viewpoint of this impersonal development in science, Alsabti, to the extent that he copied whole works, did not do anything except aid in their dissemination. He stole from the authors, but he did not steal from science, even though it's possible that some good reports were discredited by the process of his exposure. From the viewpoint of the scientific consciousness, Alsabti is less culpable than those actual scientists, currently celebrated and encouraged by the media, who have fabricated a nonexistent technique, falsified the results of an experiment, or feigned nonexistent preparations either for existential reasons or from the deforming pressure of personal ambition. The plagiarist Alsabti only popularized the truth, albeit in a despicable manner.

Moreover, the authors of *Betrayers of the Truth* are definitely wrong when they include Alsabti among the frauds and deceptions of science that not even giants of the dimensions of Newton and Dalton avoided. Alsabti actually had as little in common with science as someone who writes a bad check has with economic theory. If he had published Portuguese poems in Bengalese under his own name, he could have been a Bengalese poet, and nobody would have discovered it.

Alsabti represents a scientific horror story of entirely unscientific origin and derivation. He caused a good deal of scientific unrest, but I suspect he remained very calm himself. Like a transgenic mouse playing possum.

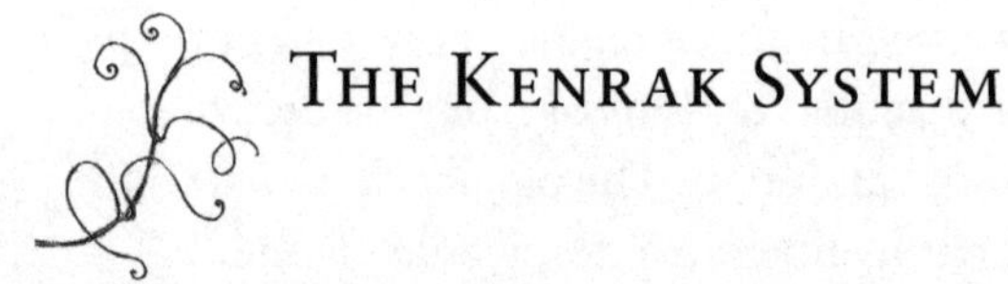

The Kenrak System

All knowledge limits the self, the German philosopher Fichte used to say.

A mere hundred and eighty years after him, a certain Zalabak, who as director of one institute of the Czechoslovak Academy of Sciences after 1970 had been "normalizing" science in the accepted communist way, developed a congenial though contradictory idea. He called on researchers to stop burying themselves in their books and instead give the self free flight, the way he did. His freest flight probably came in his directions for the treatment of cancer, which consisted of ingesting ground-up guinea pig fetuses. According to his teaching, they had to have a pH of 7 and should be taken while simultaneously crunching on carrots. It was ingenious, like the laetrile cancer cure.

The flights of director Zalabak were unrestricted as to knowledge, but alas, despite all his "normalizing" on the "scientific front," limits were eventually and unfortunately placed upon him because of his unauthorized removal of substantial amounts of building material from the institute where he worked to his private sphere. He didn't live to see any clinical proof of his unrestricted thoughts.

The unrestricted fancy of others who aren't buried in their books all the time goes right on, however, and will continue, if only because it's so much easier to take flight than to follow the state of more or less exact knowledge and acquire Fichte's insight about a thing, about its limits and possibilities, as well as about methods and their limits and possibilities.

Hence it's more than appropriate to remember, gratefully, the Kenrak system. The Kenrak system was discovered at the beginning of the 1960s under the personal supervision of the genial Kim Ir Sen and the Central Committee of the Workers Party in North Korea.

In a book on the subject, which arrived in less than ample numbers at the presidium of the Czechoslovak Academy of Sciences and which had been published thanks to the Publishing House in Foreign Languages (i.e., Russian) of the People's Democratic Republic of Korea, Pchonjan, 1965, as the title page indicates, the introduction reads as follows:

> The collective for the study of the Kenrak system, led by Professor Kim Bon Chan, discovered the substance Kenrak—a new, uniform anatomic-histological system of an organism, which is distinguished from the nervous system as well as from the blood and lymph vessels, and subsequently fully revealed the basis of the substance Kenrak; this is truly the epoch-making discovery in the development of world biology and medicine.
>
> Scientific progress in the research of the Kenrak systems opens up wide perspectives in the solution of such basic questions in current biology and medicine as cell division, metabolism, heredity, reactivity of the organism, and the origin and development of illnesses.

Let us take note of the basic features: the authors are no mere nibblers; they jump right into a new, general, unified system and its full revelation at one fell swoop. No cellular details, no biochemical details, no physiological details; they go only for the grandiose solution of what they see as essential problems of present-day biology and medicine, the solution without relations, without context, the solution *an sich*. It's something like suddenly deriving humankind from the mandrake root rather than from *Australopithecus*. This is the first constitutive trait of the Kenrak system and all systems of that kind: all at once and at one fell swoop.

Professor Kim Bon Chan is photographically depicted fixing his proud shoulder-high gaze on the not too distant future, when North Korea's Kenrak system will replace the traditional, obsolete scientific disciplines with a new, alternative science, and the People's Democratic Republic will stand at the source of the basic solutions of the problems of health and illness, cell division and heredity, all with the force of the former Michurinian biology. For the reader's moral satisfaction it's only necessary to add that Kim Bon Chan didn't allow any rival scientist to be offed, as Trofim D. Lysenko did. Or perhaps he did, but the news didn't reach Europe.

Nonetheless, in the manner of the incredible epic of T. D. Lysenko, Kim Bon Chan says further:

> In current biology a unity of an organism with surrounding environment, a regulatory mechanism ensuring the suitability of all functions of the organism, and the material basis of life events are the most important problems.
>
> In the study of these problems, biologists and doctors around the world have created a number of important theories and achieved great successes.
>
> Despite that, many questions remain as yet unanswered in contemporary biology and medicine. For instance, a speedy solution is required for serious problems such as the function and role of nucleic acid in the metabolism, the basis of heredity, the growth and treatment of malignant tumors and other diseases. . . . The biological and medical sciences [have] the task of revealing and studying new directions.

Let us take note of the vocabulary, used like a blind in a window. The material basis of life events, a number of important though unnamed theories, the unsolved question of malignant growth, "nucleic acid," "metabolism," "the basis of heredity," certain limitations of contemporary theories, and most of all "studying new directions"—all this is named on a level so general that all it has in common with the actual state of the art is the words. The writer is familiar with the matter only on the level of high school biology and is therefore completely free, at his transcendental biological level, to undertake any sort of operation: it's only a question of words, beginning words and concluding words. It's a system that, since the time of "the material basis of life events" fertilized for all time by Friedrich Engels (of Marxist fame), is used again and again to show that professional science, in this case biology, labors in a basically useless, senseless, godless, wicked, and mute way, for it simply doesn't know those words of the beginning and the end.

In this sense, the Kenrak system has timeless validity, all the more so because this system "forms the core of the Donichak teaching (Eastern medicine), the valuable spiritual heritage of our ancestors." The same blather, in the name of heritage, fills three pages of the editor's introduction, page one of the author's introduction, and four pages of the conclusions, which also appear at the end of each technical chapter: so that out of the forty-six pages of the book's text, eleven simply reiterate the statement of the epoch-making revelation of the secret of the living organism, "heretofore unknown to mankind, which is echoed widely in world biology and medicine." Here we have the third constitutive feature of the Kenrak and similar systems: in the absence of data, fill the void with babble.

There is nothing else, for the histological and histochemical pictures and descriptions show that the Kenrak system is composed of familiar textbook information about nerve endings, Kras, Meissner, and Vater-Pacini corpuscles. The Kenrak system is further composed—judging by the pictures—from nerves and torn muscle fibers and strangely cut lymphatics, as well as from primitive attempts at "introduction of radioactivity"—namely, compounds with active phosphorus—and a completely fantastic measurement of bioelectrical changes in a piece of rabbit tissue "with a single Bon Chan corpuscle." The evidence of DNA in "Bon Chan's homogeneous fluid" is something along the lines of finding evidence of Elmer's glue in the Book of Kells by touching and tasting it.

The entire technical description of the Kenrak matters is on a level that wouldn't get one into even the first rounds of a student scientific competition. Nonetheless, its creators are convinced that "this will be written in gold letters in the history of world science."

And that's the fourth constitutive feature of the Kenrak and similar systems: whether the author is a fraud or a fanatic visionary, he makes no effort to cover even the basic technical steps and forces us to accept, in the words of the good soldier Schweik, cow pies as Jericho roses.

Let us now look at the summary of these things "as yet unknown to mankind." When we read the dramatic, one-sentence paragraphs of the "general conclusion," we really have to admit that they are unknown to us, and perhaps to mankind. If it's taken as a magazine, radio, television, or newspaper report, in which the most important part, the description and analysis of actual results, is missing, and when it is read in the context of poetic resistance to the traditional sciences and in the context of citizens' experience with practical medicine, it is breathtaking. A layman, more precisely a not too sensitive layman, is apt to be deeply moved by another victory of the human spirit loosed from its fetters.

The Kenrak System

I. The Kenrak system is composed of Bon Chan corpuscles and the Bon Chan tubes.

II. Bon Chan tubes exist in two forms.

One kind travels through the cavities of the blood and lymph vessels, the other outside those vessels.

III. Bon Chan fluid circulates in the Kenrak system.

IV. Bon Chan corpuscles have their own specific bioelectrical activity.

Many experiments testify to the fact that the Kenrak electrograms reflect the overall state of the organism.

V. Large amounts of nucleic acid, specifically DNA, are contained in Bon Chan corpuscles and Bon Chan tubes.

The specific form of the existence of nucleic acid inside the Kenrak system forces us to a new understanding of the function and metabolism of this acid.

This fantastic text might well close with choruses of "Hallelujah" and become the subject of Sunday sermons by religious healers, except that it already notes how comrade Bon Chan and his collective "experience a feeling of strong excitement and wholeheartedly thank the Central Committee and our dear leader, comrade Kim Ir Sen, for the enormous care which they devoted to us and for their fervent interest in our scientific work."

Because a sufficient period has passed since August 1961 and since November 30, 1963, when these epoch-making North Korean discoveries are dated (without the Kenrak system having been mentioned ever again in scientific circles, even at the Czechoslovak Academy of Sciences or in "alternative" science journals), let us recall nostalgically the epochal revelation of the tubelets in the tube and the circulating DNA, in the knowledge that new Kenrak systems still await us before we reach the epochal understanding that there is no alternative science, just as there is no alternative poetry, and that it's good to remember what we were taught about DNA and neurotransmitters and tactile corpuscles in school, before we turn to the Kenrak manifestos, in which the world and the heavens will be new, but new only in the fashion of director Zalabak and according to the definition of the Czech cartoonist Vladimir Rencin, in whose idiosyncratic drawing a grandmother followed by a bunch of children comes out on the porch, finds the world outside covered with flowers, and calls out, "Look children, stupidity is blooming!"

In this sense even Vladimir Rencin limits the self with strong words and a strong cartoon, as Fichte limited it with his analysis of insight and his "new teachings about science," in which he didn't count on alternatives, epoch-making discoveries of Kenrak systems, or even the natural instinct of semieducated people to make asses of the rest of us.

Sukhumi; or, Recollections of the Future

One is always gratified to learn that a place one knows well has entered history. I was in Abkhazian Sukhumi thirty years before the Abkhazians and the Georgians started doing themselves in with the aid of the former Red Army. Even back then, there was a certain determinist chaos, but smoke had not yet risen from it and blood had not been spilled in it.

Down by the sea there was a summer seaside resort. Farther uphill stood a very robust statue of the Georgian Stalin, and up at the top stood the Institute for Apes and Monkeys. All around, the air was steamy and subtropical, like a laundry room.

Why I had to acquaint myself with the Abkhazian apes and monkeys has only recently become clear to me: certain militarily significant research on irradiation and infection was being carried out on them—research of which, as an "unreliable citizen" (happily), I was never a part. But I was supposed to gather certain basic monkey data: what monkeys should drink before they die of anthrax, for example.

I liked the monkeys, but not the institute. During my very first visit to see the comrade professor, we experienced a minor earthquake. Plaster showered onto the balcony, and the professor's personal monkey and I were startled—more than the professor—while I. P. Pavlov, hanging on the wall, shuddered conspicuously. When the earthquake abated, the comrade professor showed me the utopia of the baboons, who, unlike the Abkhazian and Georgian citizens, lived so freely in their vast compound that no one actually knew how many of them there were.

Then the professor showed me to the guest house in the middle of the monkey farm. My cubbyhole would have been fine. It contained, among other things, Russian curtains and a European bed. But to get into it, after 4:00 P.M., was rather difficult, because that was the time when military German shepherds were released to guard the primate concentration camp. I had to scamper from tree to tree, trying to gauge which one would be the best to spend the night on.

Following this evening scampering, a second complication would arise for me. The adjacent room in the guest house was occupied by an Armenian researcher whose life objective was to get through to Yerevan on the telephone in the corridor. He would begin calling about 11:00 P.M. and continue his efforts until something like five in the morning, when he collapsed, presumably incapacitated by vodka. His attempts to use the phone consisted of the cry "Is that Yerevan? Yerevan!" followed by inarticulate bellowing that sounded to me like "Ah-ha-eeh-ha! Ha!! Ha!!" The sounds might, of course, have meant something like "Greetings, comrade, I would like to inquire whether this is the long-distance link to Yerevan. I would like to get through to the number 85623, which is the number of my family, of whom I am at this moment thinking fondly . . ." I admired the vital energy and unfailing Ha! Ha! spirit of this Armenian researcher, but I couldn't sleep while that racket was going on. Not only because of the noise, but because one always wants to know the outcome of heroic human effort.

When I complained to Comrade Monkey Professor, he

assured me that the Prince of Oldenburg, who had founded this farm, was a great guy and never complained about anything. Instead, he had planted eucalyptuses, pine trees, and monkeys. Of the latter, only six survived the Great October Revolution and the famine in 1927. The rest had become rations for the revolutionary forces. Thus, the ways of life and all the traditions here were revolutionary. In 1927, Professor Ivanov purchased another fifteen baboons, eleven of which disappeared on the journey and two of which were turned into rations on arrival. The last two became the Adam and Eve of the Sukhumi colony. From them, seven generations had been born. The apes had been transported here thanks to an unknown KGB agent. Today there were over sixteen hundred monkeys and apes. Not counting the staff looking after them.

I was impressed, since some of the staff in attendance differed from the apes in question only to a small degree. However, I declined to continue living in the institute's guest house, saying I would hand the affair over to the Czechoslovak representatives in Moscow. To my complete surprise, this diplomatic threat worked. I was installed in the front-facing room of an international hotel in the town. Under my window was a flower bed, with a flower clock that the local pensioner would operate. And the room was luxurious during dry weather. During wet weather, a few raindrops would seep through onto the bed, but the latter could be shifted when necessary.

Now my specialist program could begin. It was fairly simple. Most of the ape research was secret. The rest of the program could be described to me, as a nonmember of the Czechoslovak Communist Party, only in a general fashion. An hour a day sufficed. During this time I learned that monkeys have thumbs, which can be opposed to other fingers; that they are, as a rule, covered with hair; and that Soviet monkeys are the best monkeys in the world. The rest of the day I spent on the beach, which was closer to the hotel than to the ape and monkey farm.

It was during these hour-long visits (in the morning, when the farm also functioned as a zoo, entry fee two rubles) that I noticed, in front of the physiology department, a stand that was brought out every morning. In it was fastened a small live rhesus monkey whose skull had been opened, the skin removed, and two electrodes inserted beneath the dura mater. He couldn't move, and as the sun traveled toward noon, he was exposed to laundry-scorching heat. With black, hate-filled eyes he observed the corpulent visitors, wearing straw hats or handkerchiefs on their heads, who arrived and departed in pure spiritual awe, as if to say, "Aha, this is science indeed. This is a sacred monkey indeed, almost like the patriarch Pimen." Some of them even crossed themselves surreptitiously.

This wasn't science, of course. This was propaganda. All that was needed to make it complete was a sign reading "Forward, Onward, with the Eighth Five-Year Plan!"

I don't know whether the little rhesus was so hardened that he suffered for the whole week. Perhaps there were seven different little rhesuses. When I complained to Comrade Monkey Professor, he told me again about the Prince of Oldenburg, who was a great guy, and of the Great Paternal War. Soon thereafter, I was sent back to Moscow.

As is evident in my other writings, I don't have old-maidish qualms about experimentation with animals. I've seen too many old maids with Alzheimer's, or with liver or kidney failure, when they become experimental animals themselves. But this was no experiment. It was an attention-getting stunt. If McDonald's used similar publicity techniques, they would have, instead of a clown, a live bull hung up in a slaughterhouse by its dislocated hind legs.

From the point of view of neurophysiology, this little

rhesus-martyr was simply an effrontery. From the point of view of history, this little rhesus was an omen that there would be bloodshed in Sukhumi in thirty years' time.

Science without a conscience is the ruin of the soul, as Rabelais wrote in his *Gargantua and Pantagruel.* Science without conscience is the death of the soul, according to Montaigne. From dilapidated souls a paranoia grows and flourishes. Its view: "Everything is always someone else's fault."

Paranoia also leads to civil wars. So I suppose the apes and monkeys have once again served as rations.

Eureka

The essence of the art of the screenplay or television script is in knowing how to present something that will be (a) accessible to people in general; (b) comprehensible even to actors; (c) manageable by technology of all grades; (d) pleasing and acceptable to producers, supervisors, and analogous organisms acting on the principle of negative creativity; and (e) fully professional, that is, resembling all previous scripts. The final product has to be obvious, graphic, pictorial, a feast for the eye and ear on the one hand and self-explanatory on the other hand, a reasonable corral for the spirit, which tends to float anywhere it wants, as the Latin saying goes. A daydreaming spirit cannot be tolerated, as it is incompatible with public enlightenment.

For example, if Archimedes were somehow to arrive for a personal appearance, direct from Syracuse and the third century B.C., and if he could be put in a studio where he was appropriately lighted and miked, with tight close-ups in bearded detail, and could then manage to state, in clear terms, his hydrostatic law on the body immersed in fluid being uplifted by a force equal to the weight of liquid pushed up by the said body, it just would not do.

The talking head, expounding on some natural law, is no feast for the eye and ear. Archimedes is no star, he's not even funny; he's a bore. The production needs action and image. It needs a professional script.

First of all, one must throw in a real Sicilian, Syracusan, street, with typical edifices of the time, picturesquely run-down, to achieve an authentic air of antiquity. Under the intense Sicilian-Syracusan sun, numbers of extras in togas fool around, along with children (belonging as a rule to members of the crew) in minitogas, and, at the proper moment, a camera assistant releases a few dogs, rented from the local animal shelter and displaying the right Sicilian meagerness of build. Authentic sweat glistens on the temples of the company members, provoking them to make their way toward a not-too-dilapidated public bath. Even the dogs are awash in sweat: artistic scriptwriters can't be bothered with the study of sweat glands.

Panning the crowd, the camera picks out Archimedes. With an abundant curly beard, he's a cross between a wise man and a wizard, and he carries in his arms an incredible load of scrolls. He has a sharp, intelligent gaze. A wizard can't just be looking around aimlessly, not even in the burning heat. Close-up of his intelligent eyes.

Deep within the eyes there appear in succession a pulley, a lever, a wheel and axle, a screwbar, a wedge, a gravity center, and the crown of King Hiero II. That's the order in which they presumably appeared in the wise mind's eye, but now they are accompanied by the music of the spheres, which modulates slowly into "The Song of the Volga Boatmen."

With this background music we see Archimedes turn to a group of lightly but decently clad slaves who are dragging an immense column or freestone, perhaps a pyramid-to-be, on rollers, with wispy-looking ropes. Archimedes' face reflects social compassion for their plight, nicely counterbalancing his implication in King Hiero's research project: the king is obsessed, for low and selfish motives, with knowing the purity of the gold in his crown. That's the cue for Archimedes to glance toward the royal residence, preferably on a hill and with a golden banner flying in the breeze. A flashback comes up inside Archimedes' head of the king brandishing the crown in one hand and a whip in the other, looking obsessed. A heavy sigh issues from Archimedes' chest, and the camera pans back to take in the whole street. Archimedes, mopping his brow as much as the armload of scrolls allows, paces, with fateful solemnity, into the public bath.

A rhythmical sound begins, wood clapping on wood, louder and louder because the decisive moment is approaching. The claps come faster and faster, like raindrops. Archimedes' leg enters the sparkling bathwater, little bubbles rise, and the leg sinks. Intermezzo with exceedingly gleaming bubbles, like a denture-care commercial or one of Cousteau's underwater films. Split-screen shot of Archimedes (from the waist up) in the bath and Hiero's crown on a shelf, waiting. Clapping intensifies. Tight shot on the edge of the water in the bath, rising as Archimedes sinks.

The camera pans in on Archimedes' forehead. Bucolic music. In the forehead, in a time-lapse sequence, we see an explosion in the Nevada desert, clipped out of a newsreel, less than ten seconds total. The music dies away, the clapping stops. Silence. Archimedes discovers Archimedes' law.

In the absolute stillness we suddenly hear Archimedes' voice—Eureka!—no big roar, just a civilized exclamation in a light baritone voice, maybe dubbed in by Fischer-Dieskau, someone like that, easy because Archimedes' lips don't move.

Jumping naked from the bath, Archimedes hops in a peculiar way, so that his left thigh and right hand, alternatively, cover the eye-attracting body parts, as the camera tracks through a rapid sequence of bath, corridor, exit, street, following the running Archimedes from behind. Out on the street the camera is already far enough away, and high enough, that not too much is visible anyway. "Eureka" resounds from near and far as the naked Archimedes runs modestly through the street, heat or no heat, dust or no dust, his shiny buttocks moving through the astonished and bath-refreshed extras, people stopping to look.

Eurekas slowly echo and fade, but reemerge in a heavenly voice, a dark tenor, maybe Placido Domingo, as the camera pans across Sicily, a few white puffy clouds, the rest of the planet (BBC, *The Living Planet*, less than twenty seconds). Patrick Stewart or some other Shakespearean voice-over proclaims that the body immersed in fluid etc. The law also appears on a parchment, or on stone tablets.

Quick cut to Archimedes in a holiday toga, the crown under his arm, solemnly pacing toward the royal residence. The slaves stop pulling and follow his steps with a spark of hope in their slavish eyes. Bach or Beethoven background music rises in volume as Archimedes goes up the hill.

Slow dissolve, obscured by sparkling water across the camera. The bubbles converge to form the words "The End."

There we have it. No one can say the show wasn't a feast for the eye and ear; as for questions of hydrostatics, well, nobody understands that stuff anyway. But nobody will ever forget Archimedes' shiny backside on the street, or the *Eureka!* with bubbles.

The real challenge to movie directors is to find a Eureka! plot for Robert Koch, Madame Curie, or Louis Pasteur. Ordinarily, science is about as picturesque as driving through a three-mile tunnel. Science badly needs script writers; alas, script writers don't need science.

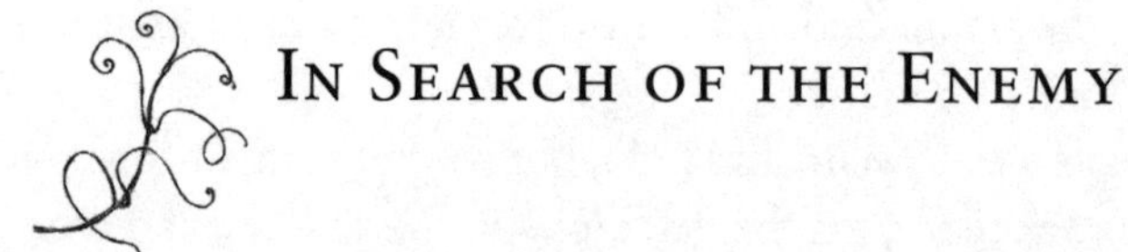

In Search of the Enemy

Not long ago, a Mayan village that had been buried in A.D. 590 by a volcanic eruption was unearthed. Archaeologists discovered, with some surprise, that ordinary Mayans were relatively well-off, had carefully cultivated gardens, woven grass mats, ornate bowls, corn cribs, and ceramic figurines. The village could afford communal workshops, storehouses, a sauna, and a shaman's office.

I wonder what an American archaeologist would make of Czech or Slovak towns and cities had a volcanic eruption occurred in, say, 1972. The gulf between the First World and the Second World is wide enough to justify such a metaphor; less developed societies are always hard to understand, since people routinely look forward and up, not downward and back.

Digging up 1972 would unearth the era of "real socialism" after the Soviet occupation, an era that has already begun to seem unreal to us, though it may still persist as metastases, deep in the body of our society.

Under the Brezhnevian ashes archaeologists would find conspicuous border fences near the border towns; the messy barracks of Soviet troops here and there, with unbelievably careless spilling of oil and negligent storing of explosives; grim, sealed-off buildings, containing secret reports of everybody and everything, in addition to clumsy bureaucratic mummies, mostly with Cretaceous intelligence; the secluded mansions of the shamans and the nobility, with ornate bowls, imported mats, and ceramic figurines; surprisingly prosperous villages with carefully cultivated gardens and supernaturally muddy roads and communal places; surprisingly well-fitted city apartments, trying to approximate Western standards; many very old cars on poorly surfaced roads, most of them leaking something; lunar landscapes in huge areas, testifying to local habits of the coal-mining industry; and a few nuclear power plants that hadn't exploded yet.

Also: dark cities, with streets deserted by 8:00 P.M. but crowded with milling shoppers in the daytime; and a good half of the workplaces empty, with a good half of the employees on the streets or in the pubs, moonlighting, visiting each other for coffee, or for vodka, or simply absent. The archaeologist would have to conclude that this society had hidden, invisible resources of some kind, and that it was inhabited by habitual vagrants, hunters and gatherers.

In reality there were very few hidden resources—maybe some clandestine arms and explosives sales. But hard currency was in any case mostly drained into Big Brother's pockets, containing and amassing funds for the anti-imperialist fight for peace, Soviet-style, as in Cuba or Angola. In reality, it was a sham economy and the workers were sham workers part of the time and private entrepreneurs, running a shadow economy, the rest of the time.

In reality it was a weird mixture of restricted civil rights and pathologically expanded personal, individualistic rights. State-supported goulash socialism was a compensation for the political disaster of 1968. It was the end of ideals, socialist ideals most of all, and the beginning of decay from inefficiency and disillusion. In the dark and cruel 1950s there had been a faction of believers, even fanatical and paranoid believers, but they vanished in the exhaust fumes from the Soviet tanks.

There was nothing left that could be "us"; there was only "them."

The "them" concept had in fact some tradition. Its existence was corroborated in the history of the present generation by "outside" dangers, so that the "them" and "us" concept was actually derived from the more concrete "outside" and "inside" concept. Human proximity was reduced to the inner circle of the family, to indoor reclusiveness, to a conspiracy of individuals who would be invisible "outside."

During the Nazi occupation and the war, it was inevitable that one would live "inside," behind the shut door, when necessary behind a curtain of running water as background noise in order to listen to the BBC or the other stations of the Allies. It was inevitable that one would be careful "outside," shunning comparative strangers not only because of the danger of expressing opinions but also, or even mainly, because of the clandestine ways one found to acquire food from relatives or acquaintances in the neighborhood or farther away.

Outdoors there were dangers—the gestapo, the collaborators, the soldiers, the dark forces listening, recording, arresting, incarcerating, killing. The shut-door syndrome developed to evade all that, and perhaps the bombs as well. I remember that during the air raids it felt safer to be at home in the cellar than in a public shelter. In fact, I am alive today just because I stayed at home one April night during a raid and did not make it to the railroad station, where I was supposed to join a team of paramedics in a well-built shelter. That shelter took a direct hit and the next morning I found there nothing but ruins and long rows of dead bodies. I never saw any of the other students from the team again.

In the two short years after the war, before the communist takeover in 1948, and in the years of the wavelike tides of liberalization, of thaws and freezes, the shut-door syndrome relaxed somewhat. Then came the Stalinist horrors of the early fifties, when the situation was analogous to Nazi times, but with a slight difference; the secret police and prosecutors and collaborators were local Czechs and Slovaks, under Soviet supervision, in some instances even former prisoners from the Nazi concentration camps, and in some instances even one's colleagues. After that "outside" period ended, people began cautiously going out again, coping with their new or old employers, businesses, factories, and unions. Somehow collective feelings developed as the waves of surrealistic collectivist propaganda receded. In public spaces the number of people increased proportionally to the decrease of official statues with socialist-realism uniforms, wearing optimistic and idiotic smiles. The 1968 disaster destroyed that episode, along with the last socialist hopes. The "outside" was back, with much more force and recklessness and with much more cynicism on both sides, indoors and outdoors, a cynicism educated by history, by the fact that "outside there" they had not learned anything from the Nazi period or the Stalinist period; in fact, they had, but they cynically pretended otherwise.

So the "them" concept was more real than at any previous time or place. It was now "everybody else," the neighbors, the officials, the authorities, the law enforcement forces, the party, the state, the Russians. The concept acquired ever larger dimensions, since at most levels, from the bottom to the top, everybody seemed to operate by the indoor-outdoor and us-them concept. Everybody would say, in private, "You see, I have 'them' on my back."

The feeling in 1970–71 was: and now I am alone again. The only collective and human proximity developed in the underground, among the dissenters. The majority were a society of defensive isolated individuals, hating most other defensive isolated individuals. Against the authorities, almost anything that did not lead to directly repressive action was okay. The rules of moral and immoral changed to the rules of visible and invisible. Artists were credible to the extent that they showed, in various disguised, ingenious metaphors, that they were "aware of all that." And the positive aim of art and literature became something like "in spite of all that . . ." Showing the incredible outside was the precondition of credibility. Viewing it from the sane outer world, one may call this a collective paranoia. In fact paranoia was the way to survive, from one's family economy to one's literary hopes. It is not that easy to label "paranoia" a fear that is a direct consequence of tough historical lessons.

It was okay to cheat "them." "They" (be it Moscow or the KGB or the state police or the politburo) wanted to be cheated; it was the only way to cope with five-year plans and pious party intentions. Officially, there was nothing wrong with that shadow economy. It was just subversive to call it by its real name.

It felt a little sinful to accomplish anything positive at all, from publishing a book (because, among other things, many could not) to building an apartment house without stealing part of the plumbing and insulation ("He who doesn't steal, robs his family"). Consequently, almost everyone felt guilty most of the time, even if it was just about the lesser guilt of someone else.

The regime underestimated the demoralization right up to the last moment. Then, in November 1989, it overestimated that same demoralization, believing it was impossible that half a million people might assemble in one place and at one time in the drive to "écraser l'infâme."

The archaeologist digging the city out of the Brezhnevian ashes would have found half the new family houses built from stolen materials; from that fact he would not be able to extrapolate that the builders might have had consciences, souls, concepts of good and evil, dignity and shame, something that could join them in a living network of common ideals.

Something like 90 percent of the population were moved to support the struggle for freedom and democracy, which had remained an unsullied childhood dream for all those forty years. But only a small percentage of them were subsequently moved to adopt new habits of work efficiency and public morality.

In my view, the problem lay in our history of negativity, from our long habit of being against things, our long-standing conviction that conditions, although unbearable, were endurable. There was plenty of evidence for "what we don't want," too little for "what we really want." There was a clear absence of concrete or positive ideals.

Philosophical essays were available, mainly in the dissident samizdat, on the meaning of November 1989, and on topics like ethics and truth. We had the moral legacy of the great Jan Patocka, and the heroic act of Jan Palach, the student who burned himself to death in 1969.

There were opportunities to show that philosophy and political thought were alive under the heavy wooden lid of official, simplistic "Marxist" leading articles in papers and magazines, occasions to show that acuity hadn't been lost in the dull times. But there was no time for broad education, there was no space in the newspapers, and there were few cultural and theoretical magazines available to the plain, bewildered citizen. Altogether there were very few realistic ideas about what people should seek in their own particular lives. Nor was there a general economic theory that gave a scenario for escape from the debris of central planning, for a capitalism that would somehow succeed without capital.

The future was defined rather as the absence of distress—in other words, the absence of communists—than as the slow birth of a new concept of citizenship. The vacuum—as it was generally felt to be—would be filled, after an interim of economic privation, more or less automatically, as a result of sound economic policy (which didn't exist) and by the inspired and enthusiastic help of the Western democracies (which turned out to be not all that enthusiastic).

The future was imagined as some sort of natural, spontaneous infection that would cross the western border, an infection by Western, mostly German, wealth, as represented by Mercedes automobiles. Western habits of work and thrift weren't part of the daydream.

The euphoria of freedom lasted about two months. During this period, the sound of the authentic human voice, represented chiefly by Václav Havel, prevailed. During the first two weeks, everybody was everybody's friend, smiling, helpful. Prague turned into an open book of poetry as young people pasted up their thoughts, their little poems of release, or quotations from Marcus Aurelius, Masaryk, and Orwell, on walls and shop windows and subway cars. After two weeks the poetry stopped, hearts closed, and little by little came the complications: the scramble for positions and incomes; people not only emerging from the underground but also coming in from the cold (even from communist circles), suddenly outfitted with Civic Forum pins on their shirts, jackets, and even their umbrellas; the founding of factions and pressure groups within the new political structures; base instincts disguising themselves as great humanitarian aims; the diversification of living standards; the frustration of travelers stranded abroad without money; the discovery that new mean personalities were replacing the old ones. All of this is well publicized, with the exception that we still have the uncorrupted voice of Václav Havel and a handful of people like him.

The arena has been well prepared by forty years (or more) of demoralization, an arena for all kinds of political and economic opportunists, cold-blooded self-interest, reckless petty larceny, crime triggered also by a broad amnesty that reflected the new moral standards of the establishment but not the pragmatic situation of the cities, and the migration of socially deprived groups from the eastern to the western part of the republic. The former official writers discovered that publishing in new private firms and selling erotic literature or romance novels is a lot more lucrative than writing socialist realist stories. The former attendants of state-owned gas stations who used to steal from the state graduated to more grandiose operations. The pub keepers who had treated their rented pubs as private property for many years already and had learned how to cheat the state and their patrons very effectively could now invest their fortunes in larger-scale enterprises. Skinheads and other groups of tough guys found a new and open playground in Prague and other cities, so that now the heirs of the German Nazis, well preserved in the German Democratic Republic, love to visit their Czech playmates.

A space opened up for international crime and international mafias. Control of these problems was complicated in the extreme by the concomitant political and moral imperative to dismantle the communist police force and create a new one, a process that led to a total crisis of identity in the police and a remarkable reduction of its efficiency. Without secret police you get freedom for everybody, from poets to cocaine pushers. The latter, of course, make better use of it. This is not an argument for secret police, just a reflection on how difficult freedom really is.

And then came a moment when the common man, home from the demonstrations of national unity on the squares and fairgrounds of the cities, asked: Who's guilty? Who did it?

The great well of antiestablishment feelings didn't begin forty or fifty years ago, when the first German soldier entered the country, but has been growing for four hundred years: ever since the hope of independence was lost, at least in the Czech and Moravian part of the country. We have, therefore, gotten used to being under somebody's boot and consequently having somebody to blame for all losses and most troubles. As the saying went in the communist era: "Bloody communists, it's raining again!"

This may just be a modern variation of bloody Hapsburgs, bloody Germans, bloody Czech radicals, bloody Catholics, bloody Protestants. It may well be that the roots of our suspicion of any establishment go back to the history of Czech and Moravian Protestantism, especially the Counter-Reformation and the Germanization periods of the seventeenth and eighteenth centuries, when the Czech and Slovak languages were almost lost because of the official German and the learned Latin. There is a long-standing habit of having overhead, in the city hall or at the castle, somebody alien, someone who talks unintelligibly or is totally untrustworthy, someone from another reality. The spirit of Kafka lived in Prague centuries before he was born. And he was there still, sixty years after his death.

True, it is impossible for anybody, aside from imbecile rightist factions, to view Havel's establishment as something alien, since its image has been formed from "below," from "among us," from inside. But it is nevertheless an establishment, and an establishment is defined, for the poor man in the street, as a priori corrupt, even if no traces of corruption exist, a condition that's possible, come to think of it, precisely and perhaps only because Havel's principled personality rules it out.

As time goes by, an impressive list of enemies has emerged, revealing the prejudices of particular mentalities and groups:

Communists

Special biological entities that never change or develop. They range from party fossils to reform committees of the sixties.

Dark Forces
Mafias formed by the former secret police. Arguably preferable as overnight entrepreneurs than as party zealots.

New Rulers
Hostility toward them fed by the traditional suspicion of intellectuals, who do indeed make up the leading force in the upper levels of the new administration. The "new rulers" make a particularly inviting target for simple minds, especially those outfitted with slogans from the extreme right.

Cults and Conspiracies
A deeply concealed conjuration of certain perennial occult forces: Freemasons, Jews, the rich, the CIA and KGB. Also scientists, because they destroy the natural world; and technocrats, because they don't adhere to the "green" ideology.

Czechs
In the "elder brother syndrome" they are still colonizers and oppressors to the Slovaks.

Other Elements
Gypsies, eastern immigrants, etc. They are either hereditary criminals or constitutively less civilized than westerners.

The greatest idiots need the greatest enemies. People who have lived through two volcanic eruptions tend to think in shorter time spans. Get the Western lifestyle as fast as possible. Get rid of all the enemies and demons immediately. We have very fast pacemakers implanted in us. We live in a constant sympathicotonia, a permanent readiness to flee or fight.

Oscar Wilde said that a man cannot be too careful in his choice of enemies. Apparently, not many Czechs or Slovaks have read Oscar Wilde.

They are in a hurry.

If there were to be another political volcano eruption now, future archaeologists would unearth hastily planted gardens, hastily stitched mats, hastily written books, and hastily engraved lists of enemies.

In Republic Square

In February 1948 I was in a crowd of students in Republic Square, protesting the communist putsch. Several files of soldiers armed with automatic rifles marched toward us from the barracks across the square. We faced them across a short space, some five or six meters. Some of them pointed their rifles toward us, others didn't. We could look into their eyes; they could look into ours. Hatred should have flashed out from eyes on both sides, but the weather was too wet. The feeling was more like a kind of repulsion, a "why do you bother us and disturb the peace" on one side and a "why are you such stupid sheep" on the other. At this point an officer roared "Disperse!" The soldiers took a step or two forward. And we began to sing the Czechoslovak anthem. The soldiers stopped, raising their guns to their chests in homage to the anthem, and the officer stopped yelling for a moment. In the cold mist the street lights had a rainbow halo, the halo of distinct embarrassment. The anthem didn't last long, and then it all began again: order to disperse, the shifting green mass with guns, the anthem. Drizzle, frost, and three nocturnal hours of a stalemate game.

We were sure that life was at stake. But we didn't know whether life was five years, forty years, or a life sentence. The soldiers knew they would get "house arrest." The communists hadn't had time yet to set up labor camps. I still don't understand why so few people at that time admitted that there were labor camps coming, along with trials and executions, that the issue was a spiritual and ideological transfer to the Asiatic steppes, that we would disappear into that metaphysical empire. Thus it was that we few students stood there alone, stepping forward and back, freezing and singing and standing our ground, all with the certainty that we had to do it.

Then each of us found his opponent. It stopped being "us" and "them" and began being "I" and "he." My "he" had very large ears and reddish hair and his coat wasn't buttoned to the neck. Maybe he got dressed too quickly, or too unwillingly. Instinctively—or maybe I read it somewhere—I knew that a soldier with a not-quite-fully-buttoned coat was not 100 percent a soldier. In a way, this became my own personal hope. When the anthem was being sung, I sang it to him. He stared, hesitant. One finger on his glove was frayed, and his ears were red. It was just as terrible and embarrassing for him as it was for me, but to bolster his confidence he stared at me as if I were a two-headed calf.

And then I realized that he was Ginger from a French short story I had read with my mother at a time when it was believed that every language you knew extended your humanity.

In the story, Ginger caught a mole and didn't know what to do with it. He dropped the mole and it squashed blackly, squeaked softly, and did not move. Ginger picked it up, but the mole still did not move. Its paws were spread. Ginger was scared. He had not wanted to hurt the mole. He just didn't know what to do with it. The mole had ruined itself, in fact it had done that on purpose, to make trouble for Ginger. The mole was guilty. Ginger threw the mole up and the mole fell to the ground and still did not move.

Ginger threw the mole way up in the air and it dropped with a thud. Blood squirted from the pink mouth lined with tiny teeth. The black velvet fur got sticky. The mole was dead and it was more than ugly, it was an insult.

Ginger had tears in his eyes—maybe of repulsion, maybe sadness, maybe sympathy—and he picked the mole up and threw it up with all his strength. The mole fell back down and its black fur split open. Flesh and juices and clotting blood showed.

Ginger picked it up again, weeping, and threw it up with all his strength, desperately, way up into the sky, hoping it might stay there. It fell down. It was a bloody, twitching, disgusting, and hostile alien thing.

Ginger cried and kept throwing. The dead mole turned into red primary matter. Ginger had blood on his hands and blood on his coat and blood on his face, and he kept throwing, and the repulsive blood kept falling down.

It was something like Flaubert's story of Saint Julian the Hospitaler who started by killing a mouse in church and went on to slaughter forest and field animals, then became a successful warrior and eventually killed his own parents by mistake, after which he became a saint, keeping the dying warm with his own body and eventually coming face to face with Jesus. But first he hit that mouse under the altar.

For Ginger the throwing became a defense against repulsion, and repulsion was a defense against the original sin. If Cain killed Abel, it was because Abel became repulsive to him after the first blow.

When we hit, slap, cut, and beat we expect an aesthetically acceptable response, as in the movies. A response that includes collapsing, yelling, smelling, screaming, and shedding blood is an unacceptable personal aggression, and revenge must be taken, by dealing more blows.

First there is the motive; then there is the first blow. And then there is blood, and it goes on and on. Adrenaline, sympathicotonia, an inner alarm leading to an animal attack, often without animal taboos. Such is the order of the child versus the mole, such is the order of the soldier or policeman versus the chickenshit intellectual. Such is the order of the white man versus the colored man (and sometimes vice versa), of the landlord against the bothersome immigrant.

And such is the order of the individual of a simple mind against a phenomenon that is too complicated. There are many kinds of human hatred. But the hatred of stupidity and ignorance of the more educated is one of the strongest because it is one of the most well disguised. It is entirely apt that what happened after the event in the square in February 1948 and lasted for another forty-two years was officially called a "class" struggle, which Czech humor interpreted as the struggle of those having finished five classes (school years) against those having completed twelve of them.

We were not yet in a real class struggle that day in Republic Square; the end of the war, with its abundance of beating, yelling, bloodshed, and dying was still close. It must have had a dampening effect, and that is why we avoided what happened to students at the opposite pole of that historical episode in November 1989, on National Avenue. A lot of police "Gingers" had their fun, and we witnessed what Konrad Lorenz calls emotional infantility, what seems to be part of the training in courses for special commandos or terrorists.

Training like that takes place everywhere in the world, on the right and on the left, in the interest of the state or in the interests of some emotionally infantile lout with five years of school or less. The goal of the training is always reached, be it in football crowds or as a supposed defense against niggers and foreigners and Jews. "The asylum seekers are not the cause," as *Stern* magazine correctly commented about recent events in Germany. "The asylum seekers are a pretext."

The training results in what Lorenz described with the formidable phrase "thermal death of feeling." That is, "irresponsible and infantile striving for the immediate satisfaction of primitive wishes," leading to "the absence of any responsibility and respect for the feelings of others." Others, whose only guilt sometimes is that they do not think like us.

It isn't just a group lapse into a more primitive stage of human evolution during the permanent human attempt to achieve more consistent humanity; it's an active group protest against an anthropogenesis that leads to a differentiation too complicated to be solved by a generally comprehensible formula or an equation. It is the reductionism of a simple, sometimes stupefied mind. It is a relapse, excuse me, to the Aztecs, for whom bloodshed guaranteed the future duration of the cosmos.

Erich Fromm wrote about the preindividual state of human existence. One cannot help feeling that individualism, as fostered by the Western form of civilization, is not so suspect as some intellectual critiques have made it sound. The degeneration of Western citizens into states of emotional infantility and thermal death of feeling is even more to be feared, and it is detestable.

According to Fromm, this preindividual state includes bloodthirstiness, killing from passion, killing to suppress other forms of life and in order to gain life, since the strength for hominization is still incomplete. It is the attempt to transcend life at the level of deepest regression into prehuman primitivism. A life balance, in this archaic sense, is achieved by killing so many others that we ourselves—satiated—reach the state of reconciliation with our own death, a state of readiness to sacrifice ourselves, Fromm says.

I have no doubt that if the rules and laws of civilized countries were less efficient, our trained primitives would readily regress to this level of bloodshed from passion, to this cycle of killing in order to transcend one's own life and death.

I have never quite rid myself of the notion that I'm still in

Republic Square, that I face my Ginger, that I am a mole, a substrate of the class struggle. Some of our radical purgers of communism resemble not only those officers who yelled "Disperse!" but also those intellectuals who informed on everyone in Republic Square and who welcomed labor camps for the class enemy.

I have never believed that regression to emotional infantility is solely a group phenomenon. It only takes place as a physical action in terms of a group. On the level of the so-called spiritual life, thermal death of feeling is quite a common phenomenon.

So there I stand in Republic Square, singing the anthem from time to time. And I am freezing. It's terrible and it's embarrassing.

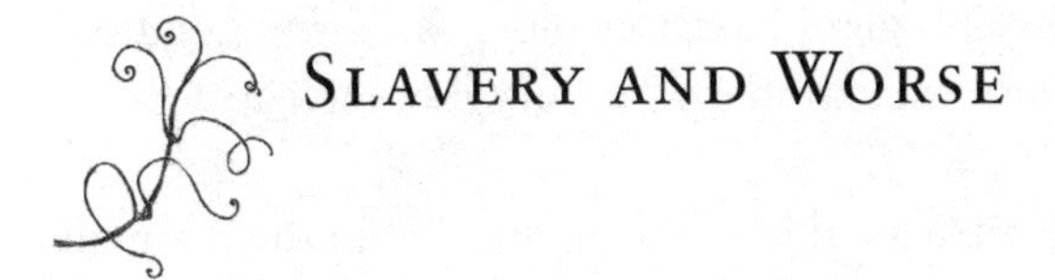

Slavery and Worse

In a side room of the Pergamon Museum in Berlin I discovered a stone tablet on a wall. It was an edict, very probably Diocletian's, that had been displayed in the marketplace at Aizanoi. It gave the binding prices of commodities in denarii, in relation to the price of gold, one pound of gold equaling 72,000 denarii. Livestock prices included the following: cow *(formae primae)*—2,000 denarii; goat *(formae primae)*—600 denarii; *dromedarius optimus*—20,000; two-humped camel—60,000; human male from sixteen to forty years—30,000; female in the same age span—25,000; male over sixty or under eight—15,000; female same age—10,000; riding horse (*equus currulis*)—100,000; military horse *(formae primae)*—36,000.

It was very reassuring to see these fixed prices in our realm of vague values. Harold Morowitz estimated in the mid-1970s that the simple chemical materials making up our body were worth 97 cents, but that if you calculated, more properly, the value of regulatory, signaling, and effector biochemicals such as enzymes, hormones, nucleic acids, hemoglobin, contractile proteins, albumins, globulins, hyaluronic acids, collagens (not to mention interleukins and endorphins), you might come up with $6,000,015.44 for a man weighing 168 pounds.

If we take the present price of gold as $350 per troy ounce, the biochemically defined male slave in prime age and weighing

168 pounds would cost around $275,000 in a hypothetical market, the equivalent of about 4 million third-century denarii. Not a tremendous difference, taking into account that the six-million-dollar man might have lots more skills, imagination, information, and education than a Phrygian slave in a Roman market.

This is of course an abhorrent calculation in the realm of human values, 315 years after the Habeas Corpus Act. Even if the price of all human organs that may be used for transplantation is about $80,000 and the highest life insurance paid may be somewhere around $1.5 million, just about 5.5 times higher than the hypothetical current price of a 168-pound slave under the age of sixty.

And are we really in the twentieth century in all locations on the planet? One of the troubles on the spaceship is that we not only live in different places, we also live in different times. And we always have. The peculiar trait of humankind is that almost nothing is ever really solved. Nothing in the history of life and in human history ever ends. We are no longer in the Stone Age, but there are still secluded human groups that use only stone tools. We are no longer in the Middle Ages, but we have a dominant Middle Ages parasitic disease like malaria, which killed more people—2 million—in 1991 than in any medieval year. Tuberculosis, the eighteenth- and nineteenth-century plague of Europe, threatens more people in the Third World than AIDS, cholera, dengue, and other infections combined. We still have social and racial ghettos. We have pre-Columbian myths prevalent in very post-Columbian Americans, and twelve centuries after the Mayan priest-astronomers met in the holy city of Copán to synchronize the two calendars, the time of gods and the time of man, we live even numerically in different years and centuries of different profane and sacred traditions.

Joseph and his brothers, sons of Jacob, were sold in biblical times to Potiphar, the Egyptian. Biblical times persist, not only in terms of human and divine values. Slaves no longer form a vast majority of the population as they did in ancient Rome or Greece. The last legal slavery was abolished in the 1960s in Saudi Arabia, and in 1975 or even later among Berber Tuaregs in the Sahara, but there are still slaves in some dark spots on the planet, like the border between Angola and Zaire, southern Ethiopia and Sudan, in the southern part of the Arabic peninsula, and maybe in some remote corners of Southeast Asia and Melanesia where the worst kind of slavery, feeding captured people for slaughter and cannibalism, was recorded within the past hundred years. A similar report came out of civil-war-torn Angola within the last ten years.

There must still be millions of slaves around, if you take the definition that slaves are not legal subjects but legal objects, people who are the property of other people.

We have slaves with nicer names but in more desperate conditions. There are indentured servants, debt slaves, child slaves, and peons and illegal immigrants who are more cruelly exploited at times than were the slaves in Rome or China. Slaves in the old days were regarded as a kind of investment, a long-term staple commodity whose value would drop after maltreatment or be lost if they were killed. Today, there is only a short-term interest in the survival and health of indentured servants and purchased child prostitutes or drug traffickers.

Fewer slaves are generated now by slave raiding and by capture in war, but there are still clandestine slave markets. People even sell themselves. There are sales of women and children, as in China with its successful population-control policy.

And how could one even compare the treatment of an ancient Chinese concubine slave to the present-day trading in white-skinned and brown-skinned teenage girls from Bangladesh, India, or Eastern Europe who are sold to brothels from Arabia to Germany? Among our abducted girls there has been no known example like that of Malinche, the multilingual Aztec girl of noble origin, reduced to slavery by family rivalries and given by the chiefs of Tabasco to Cortés. Malinche had a significant political role, and she gave Cortés his son Martín.

Can the powerlessness of a slave be compared to the powerlessness of children anywhere? In the poverty-ridden areas of northeast Brazil and northern Thailand, "entire villages are bereft of teenagers," I read in *World Watch* magazine. "Some have ended up in the brothels of Bangkok and Ranong, locked in tiny cement cubicles, servicing 10 to 15 diseased clients every day. Some have ended up in Rio, bought by wealthy ranchers who gang rape them to death in a regular Saturday night ritual. . . . Today, children are being bought, sold and traded like any other mass-produced good. In the ever expanding market, child prostitutes are among the hottest commodities." Two hundred and fifty thousand to half a million children under sixteen are involved in the Brazilian sex trade. In a recent brochure a Dutch travel agency described the Thailand prostitutes as "little slaves who give real Thai warmth."

How could one even compare the lot of a 110,000 denarii slave family in Diocletian's time, or in ancient Egypt, where the smarter ones could achieve even high bureaucratic or scholarly positions, to the situation of the 18 million refugees from famine and war areas in Africa and Asia nowadays who can only beg and die? By chance, the same figure, 18 million, is the number of Africans delivered into the Islamic trans-Saharan and Indian Ocean slave trades between 659 and 1905.

The status of slaves in the Ottoman Empire, from the janissary soldier to the harem girl, and even in the Caribbean plantations of the sixteenth to the nineteenth centuries, was not worse than the status of people in the Gulag from 1920 to 1956. About 15 million people died in the NKVD and GPU camps, and 6 million died as serfs in Ukraine and North Caucasus from famine induced by forced collectivization in 1932-33 alone. There was no recorded cannibalism among the slaves working the Laurium silver mines in fifth- to third-century B.C. Athens. There were about eight thousand proven cases of cannibalism in the Volga region in the 1920s.

It is unlikely that the female slave sold in Aizanoi for 25,000 denarii was mutilated. About 80 million girls and women living today in sub-Saharan Africa, the Arab world, Malaysia, and Indonesia (not to mention migrant populations even in Great Britain where some 10,000 girls are at risk) have undergone or will undergo female genital mutilation due to religious, traditional, ritual, and other "natural" reasons in the context of indigenous societies and their paradigms of illiteracy and superstition, which we modern folk broadmindedly "accept" as the "values" of non-Western cultures.

Female genital mutilation is not one of the horrors of slavery. It is a cruel session of torture one would not perform on a woman slave, since the shock, infections, injuries to adjacent organs, difficult labor, and risk of infertility would decrease her value to well below 25,000 denarii.

Even self-sold indentured servants in modern sports may achieve, under good totalitarian conditions, a status comparable to that of the Roman gladiators before Spartacus, rich slaves without any human rights. There is the story of the Soviet soccer team Zarja Voroshilovgrad in the 1970s. The party and the city bosses decided that they would propel the team into the first division by means of an efficient Soviet amateurism. The players' salaries were increased and they were forced to marry local prostitutes. They had to spend most of their time in their training facility outside the city, however, which left the ladies at the disposal of the party and city bosses and allowed the players to save their vital forces for performance on the field. The highest hopes were pinned on an excellent striker who was made to sign with Zarja in return for the promise that his brother, who was awaiting execution somewhere in Doneck, would be transferred to a Voroshilovgrad court for another trial (although he never did anything in the Voroshilovgrad area). In spite of all these measures, Zarja stayed in the first division for only one year. The brother was sent back to Doneck and executed. I don't know what happened to the prostitutes, but otherwise the story sounds like Seven against Thebes, including Polynices and Eteocles. Without Antigone.

In our contemporaneous mix of centuries we tend to abolish words rather than facts of human suffering and despair. We tend to take changed statistical probabilities for moral victories.

Red Noodles; or, About Uselessness

Scientific research institutes in the communist era were usually surrounded by fences and guarded by gatekeepers, sworn to their posts, who took care to ensure that not even one scientific research mouse escaped. In front of one such scientific research workplace, not far from the capital, near the fence and not far from the gatekeeper, a completely ordinary taxi stopped one day, containing one middle-aged taxi driver, one old woman, and a milk can filled with milk. The old woman had, around her neck, something like a fox stole that had gone through a great number of disciplinary proceedings.

At that moment the gatekeeper happened to be giving a hard time to one of the scientific research workers who, on that day, hadn't yet completed the formalities connected with entry to his workplace, and who therefore was still proving he was who he was.

The taxi driver stepped up and asked, "Do you test milk here?" The scientific research worker, urged on by the gatekeeper, answered, "No, indeed. Who gave you that idea?"

"They sent us here from the hospital," said the taxi driver. "From the hospital in Krc, where they sent us from Charles Place. And before that they'd sent us to Charles Place from Bulovka hospital. And they'd sent us to Bulovka hospital from

Klimentská Street." He was a conscientious taxi driver. "It comes to seventy crowns already," he added.

"And what's with this milk?" asked the gatekeeper.

"The milk is poisoned," said the taxi driver sadly. "Come ask this lady."

So they did.

The old woman explained in a trembling voice that the milk was poisoned because the gang that lived with her wanted to poison her; they'd also given her a bitter apple casserole and red noodles. All of it poisoned. Mixing into her speech various oaths from the early days of Christianity, the old woman pleaded most pathetically to have the milk tested.

"But madam," said the scientific research worker, "We can't do that, we don't have anything for that here. There are special laboratories for that."

The old woman cast him a reproachful look, insofar as an old woman's watery eyes, covered with wrinkled lids, can cast such a look, at the white coat and at the institute's extensive buildings and said, "But they're giving me red noodles. And this soup that makes me throw up ever since my husband died; he had gone up to the attic, where it was leaking and he was digging there, to drain the water away, but there was such a big puddle that formed on the concrete that it wasn't possible and so they're giving me poison now so they can have the house, ever since that time when there was that puddle, and my husband, who has died, went up there, so they're giving me the red noodles, doctor, for God's sake I beg you, you can't eat that stuff, I drank a little of the milk and came here right away all the way from Brandys, they're all against me there and that's why they're giving me that milk and those noodles."

The scientific research worker glanced at his watch and went off to make some phone calls.

He learned—after several fruitless inquiries and wrong connections—that in the local toxicological laboratory they weren't allowed to accept red noodles or undrinkable milk unless it was accompanied by a recommendation from a treating physician or the local police. It wasn't done for private individuals.

"But the doctor's afraid," said the old woman, "and the police are on their side; they've got contacts everywhere. I've been there lots of times already so for God's sake, I beg you, doctor, if you could check whether the poison's in there, I don't have the noodles but this is the milk." She held up the milk can. The taxi driver and the gatekeeper nodded their head. That was the milk.

"So where are we supposed to go?" asked the taxi driver. "I'm not going to charge her any more," he added, with a glance at the fox in which the old woman was dejectedly slumped. The fox looked more depressed than ever. The old woman repeated the whole story from the noodles and the puddle and the husband to the gang's contact in high places. "That," said the scientific research worker, "I don't know about." He really didn't know. The time allotted to scientific research work was passing. The gatekeeper and the taxi driver looked on with concern, albeit uncomprehendingly. It was clear to the scientific research worker that the old woman was a little *too* old, and, as sometimes happens, somewhat disoriented. That because of this, and maybe rightly, nobody at home believed her. That there was an insignificant percentage of a chance that the red noodles and the milk could really have been poisoned. That his specialization and his scheduled tasks, whose allotted time was rapidly passing away, had nothing in common with noodles and old women in mangy fox stoles. That it was, so to speak, useless.

Nevertheless, after a while, with a resigned glance at his watch, the scientific research worker stepped into the taxi where the sickly fox dragged its tail in the poisoned milk and, as the taxi driver settled in, said to him, "Let's go."

The taxi drove off into the hazy autumn countryside and disappeared, jumping over the small, freshly repaired potholes of the E-4 road. The gatekeeper closed his gatehouse, filled with the characteristic aromas of soup eating and pipe smoking. What should the research worker have done? Intercede at a police station outside the woman's district, or take her to a psychiatric clinic, in either case knowing that the old woman's dispute would not be resolved?

I suppose that his act wasn't entirely sensible, but it was human.

It served no purpose, but it was human.

It was pointless, but it was human.

In fact, the more useless, the more human.

Perhaps there's a law in the world, a law that we discover toward the end of our days, that even the useless has a certain weight.

The taxi, meter turned off, jolted into the city center.

The old woman said, "And, doctor, even that apple casserole was sort of purple, and my husband when he was digging in the attic caught a bad cold there, and they've got contacts at city hall. So please, doctor, tell them, where we're going, to take the milk off the bottom. That's where most of the stuff is."

The taxi driver nodded, to indicate agreement.

The scientific research worker's watch stopped. He felt very sad. They drove on, carefully yielding the right of way to vehicles on the main road.

No

Everyone has three or four poems that he remembers for unknown reasons. One of mine is by Jacques Prévert. At least, I think it is Prévert; I read it in 1947 in a Czech translation, and since then I have been unable to find it anywhere, except in my head. It is called "Poem on Disagreement" and it goes, in my retranslation, something like this: From time to time, one must say No. No, mother, no, father, No, Mr. Professor. No, just a little bit, For the sake of scarce honor.

One can't always say what is so attractive about the poem that has imprinted itself in one's brain. Thinking about this one, I would suggest that I especially like the "little bit" and the "honor," the idea of human dignity. I don't like obsessive, neurotic noes; they belong to the Goethe spirits who "steadily negate," who play devils in run-of-the-mill Faust performances. I prefer the little, calm, nondevilish no in a concrete situation, provided one is reasonably sure about its unacceptability. With such a small, concrete no, one gets out of the collective of all-yes or all-no, one gets off alone. This is an important aspect of dignity, of sanity and maturity and freedom. Certainty is subjugation, remarked the French biologist Jean Rostand. Maturity means ad hoc noes to the tradition, to probability, to juvenile negativism and inveterate opportunism. The small no is self-defense of the human mind.

"Care about the symmetry of your cross," wrote the Polish aphorist S. J. Lec, whose volumes of *Unkempt Thoughts* might be called no-books.

The obligation is to say no not when it is obvious, but when it is unexpected. Not no to an authority as such, but no to a wrongful authority, so wrong it may be surprised by it. "Satires which can be understood by a censor are rightly forbidden," said Karl Kraus.

Under totalitarian regimes, we had advanced courses on why, where, and when to say no. In the long run, some noes turn into matches starting little fires or small explosions. And most noes start snowballing, sooner or later. No no is wasted, in the sense of the Latin proverb *Gutta cavat lapidem non vi sed saepe cadendo* (the drop excavates the stone not by force but by falling continuously). The lie is fearful. The lie can't afford a defeat, but truth can, said Rabindranath Tagore. I used to know some constructive types, building concentration camps, and some destructive types, demolishing them (Lec).

I remember a representative, seemingly negligible but meaningful no situation on Prague's Wenceslas Square, which appears to be the locale of choice for most Czech no events in this century. In the dark seventies, there used to be a popular market, a sort of informal stock exchange, of badges and pins there. It was a time when a plainclothes policeman was found in any public place where more than four people had assembled.

And here was a tight, compressed group of about twenty men, some clearly normal, some suspicious-looking, with the sort of Brezhnevian faces that dominated all the party and police ranks. All with burning eyes, watching the few lucky ones who got into the center space and could reach for their wallets and display their contents. The wallets displayed in the center of the group were all of fine leather, and they contained black or maroon or dark blue velour cushions on which the treasures rested: colorful badges of Botafogo Rio de Janeiro, Estudiantes, FC Barcelona, the All England Tennis Club, the Los Angeles Lakers, Sampdoria Genua, Bayern München, Arsenal London, Ascot Racing, the New York Giants, Rapid Wien, Real Madrid, the Toronto Maple Leafs, Ajax Amsterdam, the Cleveland Indians, Tower of London, Olympiacos Pireus, Federbahce Istanbul. Business was lively. One could get about four Bayerns for one Los Angeles Kings, and one Cleveland Indians for 9.5 Galleria Uffizi, which was a little enigmatic because nobody even knew whether they played baseball or rugby. Everyone was totally absorbed by the trading and by the beautiful little enameled badges and pins.

And then came a little man, wearing a typical Russian velour hat and unbelievably baggy pants. He squeezed his way toward the center. His eyes were burning too; he was obviouly excited by the prospect of trading because nobody back home would have the least idea of Federbahce, Ascot, or the L.A. Lakers, nobody in Moscow or Irkutsk or Sverdlovsk . . . Being small, he succeeded in worming his way finally to the center, where he reached for his wallet and displayed, with shaking hands and immense, childish hope, his treasures. They were pinned on gray cardboard. And they were: Lenin in Leningrad, Lenin in Finland, Lenin with a Red Banner, Lenin with a Cap, Lenin with Marx-Engels-Stalin, Lenin Standard, Lenin Deluxe, Lenin Combi, and Lenin Four-Wheel Drive.

Obviously he expected to get at least one Estudiantes for twenty-three Lenins. But once his wallet was displayed, a tremor of disgust ran through the crowd and the nearest Czech collector told him: no.

No. Don't bother us. You can't get anything here for these . . .

The poor Russian looked around, slowly closing his wallet with a gesture reminiscent of the second entombment of Lenin.

All the faces, normal or Brezhnevian, signaled a clear and definitive no. Get out of here. Fuck off. He disappeared, smaller than before in his velour hat and broad trousers. Never before had I felt pity for a Russian, but I felt it for him. Him with his Lenins. He was, in spite of the hundreds of KGB men in Prague and half a million Russian troops around, a minority, a poor little man destroyed by a clear-cut and concrete no, shared by all twenty specialist/collectors, including the Brezhnevian ones. I felt pity for the man, but I approved of the no.

There was no moral ethos in the no. But it was one of the twenty or so shades of no that preceded the collapse of the regime. The badge and pin collectors' stock exchange no may eventually prove to be as important as the philosophers' no.

It was not just a clash of mean or childish interests. It was a clash of two kinds of game, in a subtle way a clash of two kinds of culture. It was a European no to the belated Byzantium. The situation was an image of our basic Czech assumptions, formulated by S. J. Lec: all gods used to be immortal.

A system based on the "class" struggle, that is, the struggle of those who have had six years of school against the ones who have had twelve, is an ideal device for turning remedies into poisons, corrections into disasters, small events into epileptic seizures.

The good soldier Schweik was well aware of this, and he remains an inspiration in all kinds of pressure situations. His way of saying no by saying yes is very effective in epileptic systems. It is even a moral form of behavior provided one is sure that one stands with the oppressed and against the oppressors.

Schweik is not only a typical Czech character. He is essentially a product of middle European geology, although I know of some Schweiks who were of quite different ethnic origins, and who were simply adapting to the geology.

My best example is Professor Cort, a Jewish Polish-Canadian, who for some very pragmatic reasons immigrated to Prague. He established a first-class laboratory for kidney physiology. In addition, he was the obvious choice to translate scientific texts into English.

Thus he was asked, in the dark seventies, to provide the English version of the annual report of the Institute for Clinical and Experimental Medicine. The scientifically less-than-impressive report was introduced by the windbaggy minister for health. One problem was that the introduction was stupid even for a minister, so that Dr. Cort suffered. The text also contained a quotation from Marx that was so distorted that it could have been by Saint Augustine. Cort decided to improve the text by replacing the quotation with a genuine paragraph from Marx on public services and human dignity.

The result was a much better version of the introduction. In a way it was a yes—to a socialist introduction by a socialist minister to a socialist annual report—and I am sure Cort knew what he was doing and what he could probably expect by way of feedback, but he simply couldn't stomach a mangled quotation.

The report was supposedly proofread by all authors in all language versions, appeared in due course, and all remained quiet. Because who reads socialist annual reports? But then, after a year, the epileptic seizure occurred. A Vietnamese post-doc in surgery, sitting somewhere in the institute for a year and realizing that he might never learn anything (or need to know anything) about open-heart surgery, decided he would at least learn English. He did it in a peculiar Vietnamese fashion, through comparative reading of the Russian, Czech, and English versions of the institute's annual report. After three months of deep, basically Zen immersion in the texts, he discovered the discrepancy of the Marx quotation in the English version, as opposed to the Russian and Czech versions.

He was sure—well aware that he had been sent to a country where Lenins had a lower exchange value than New York Giants—that he had uncovered the creeping counterrevolution, and he decided to ornament his surgical curriculum with a political one: he reported his discovery to Hanoi, Hanoi reported it to Moscow, and Moscow insisted that the Prague revisionist be revealed and made an example of. The issue wasn't the real Marx quotation, but the difference between Russian (language of communism) and English (language of capitalist imperialism).

Of course the minister of health denied that he ever read the proofs, though he had signed them. And of course the genuine Marx paragraph was taken to be a Jewish/bourgeois/nationalist deviation. Of course Cort was to blame, although he had simply improved the text above the level of ministerial feeblemindedness. He finally had to leave the country, despite the fact that he would have been more useful in Czechoslovakia than in the United States or Canada.

But at least he left behind the inspiring, inconspicuous, Schweikian no.

There is another kind of no, perhaps the most important and essential one: a no that is not spoken, just performed.

I did my medical studies in Prague in the rough years of flourishing and later, declining Lysenkoism and the more durable Pavlovian medicine and "nervism" of Academician Bykov. The main argument in favor of these ideas was that they were Russian and Russian was superior to anything in the world. Who invented television? This was a typical Czech joke of the period. It was I. P. Televisorov; redecorating his apartment, he carried pictures from room to room.

I was practicing in a clinic for gynecology and obstetrics. The full professor was a dedicated Pavlovian: birth had to proceed by purely natural means. No hormones, no drugs, no surgical interventions, just conditioned reflexes, deep breathing, and lots of confidence. In our ward, we had a nice young lady, full of confidence—she was the wife of a policeman—who had been breathing deeply since the middle of the morning. The pains came and subsided, came again; the cervix did not dilate properly, and she was totally exhausted.

The professor's sermons on natural childbirth, the greatness of Pavlov, and the importance of a strong will only made everything worse. He seemed to blame the poor red-haired girl, pale and sweating, for this utterly unnatural deviation from politically correct childbirth. He took most of the medical documentation into his office, locked it, and left, disappointed, in the late afternoon, confident that nature would take its course sooner or later.

We felt deep pity for the poor girl, who was really very nice, with freckles on her little nose. But only the senior staff person on night duty, Dr. Cech, could make a decision. He decided no, without actually saying the word. He simply told us to get scrubbed up and prepare for a cesarean. He asked only us students, not wanting to interfere with the careers of the staff physicians. It was breaking the rules, the discipline, the professor's will, and nature's way. I was picked as the first assistant for the surgery, maybe because I forgot to look alarmed, which would have been because I was utterly alarmed. It was the first surgery of my life. I was holding the hooks, dizzy and sweating as much as the poor young mother who disappeared under the green drapes. It was midnight, the operating theater was deserted, sequestered from political correctness, medical documentation, and institutional discipline. All Dr. Cech and the rest of us had in mind was to save the girl from another round of suffering. We had no excuses except humanitarian ones, politically suspect.

I would never have believed how much blood, sweat, swearing, sharp metal, and rude force humanitarian motives required. All went well, thanks to the skilled nurses on hand. It was twins, the second one a real surprise, each of them at six pounds, both happily and vigorously screaming.

In the morning the mother was happy as a sunflower and thanked everybody around, including the disappointed full professor, who had to be told that there had been an imminent danger to the babies. The red-haired mother never realized that it was really the tact of Dr. Cech that was to be thanked. Many times since I have thought of the policeman's twins, wondering whether they too have learned something about the vital necessity of noes. And I remember Dr. Cech, whose name means Czech, by the way. We are not a nonheroic nation, Karel Capek stated in 1924, it's just that we have a tendency to apologize for the fact that we allow ourselves, now and then, to be courageous.

The clandestine, persistent, inconspicuous noes of small nations have helped their survival and still function like oxygen during the most life-threatening moments. These noes to gods, rulers, and dogmas help such nations keep their identities clear. Jan Werich, a great Czech comedian, coined a typical sentence that used to be, together with Werich's portrait smoking a cigar, pasted on the wall of almost every laboratory, doctor's office, and even in those plain municipal offices not shared with plainclothesmen: "A man should be what he is, and not what he is not." It was an unobjectionable sentence, but everyone understood what Werich meant by "what he is not." It was a very broad no that survives into the present, a logical no even under democracy and most certainly in biology.

The immune system of the body is a device for turning yeses (to functional states and organizations of the body's own cells and tissues) into noes (to alien molecules and cells). Nerve impulses are sweeping yes and no changes along the nerve fiber; the opposite regulatory functions of the two autonomic nervous systems may be, again, poetically designated as yeses and noes, as can all the intercellular and intersystem signaling in the body, with noes, negative feedbacks, essential to the basic order and maintenance of the organism.

The hominization process implied a long-lasting, tacit no to instinctive forms of behavior, a forceful no in spite of the hundreds of millions of years of success for the instincts.

The vertebrate evolution began as a haphazard no, by one or a few species of primitive chordates, to the Cambrian mass extinctions of half a billion years ago.

The multicellular organisms started with a no by unicellulars to life conditions, such as scarcity of food; even now, some amoebas produce a chemical signal that induces some fifty to a hundred thousand individuals to form a colony and, later, an almost integrated organism of two or more differentiated tissue types, all within twenty-four hours.

And, to get some concrete poetry into the story, one of the most potent and ubiquitous noes of the cell or tissue to intruding enemies happens to be NO, nitric oxide, the molecule of the year in 1992, as nominated by *Science* magazine. NO, a real surprise in biomedical research, a small, short-lived molecule passing easily through cellular membranes, turns out to be a most potent defensive weapon and signal among cells.

A global no to life in a methane and ammonia atmosphere, uttered by the first photosynthetic organisms producing oxygen, one of the most life-endangering elements, changed the basic energy metabolism of life and produced chains of noes to free oxygen radicals in the living bodies that, ironically, make successful use of these free radicals in the warfare against alien intruders.

Evolution is in one basic respect a no to conservation, to the imperfection of conservative mechanisms of matter: life learned how to profit from this imperfection, and it represents the only cosmically successful no to stability—and to eternity, if you like.

We are here thanks to an immense series of noes.

I don't know whether the final answer is a yes or a no. Actually, I don't know what the question is. I was aboard an airplane once that caught fire some two hundred miles away from Kennedy Airport. The alarm sounded in the cabin and in the cockpit, and amid the screams, cries, smoke, and nervous orders, the oxygen masks dropped. My first thought was, Yes, that's it. My second and final thought after a half-second was, No, that can't be it. A vigorous, silent no. And that no has been final, so far.

Notes with Selected Bibliography

This Long Disease

5 Friedrich Wilhelm von Schelling (1775–1854), German philosopher.

9 Peter Medawar, *Future of Man* (New York: Mentor Books, 1967).

10 Susan Sontag, *Illness as Metaphor* (New York: Farrar, Straus and Giroux, 1978).

11 Stanislav Andreski, *Syphilis, Puritanism and Witch Hunts* (Basingstoke: Macmillan Press, 1989).

12 Leon Daudet (1867–1942), *Devant la Douleur* (Paris: Nouvelle librairie nationale, 1915). Passage translated by Miroslav Holub.

Kidneys and History

13 Milan Kundera's comment on Tycho Brahe is from *Immortality* (New York: Grove Weidenfeld, 1991).

14 Blaise Pascal (1623–1663), French mathematician and philosopher. *Pensées* was published posthumously in 1670. Passage translated by Brigitte Frase.

The Experiment of 1668

27 Jan Babtista van Helmont (1580–1644), Belgian alchemist and chemist, first chemist to recognize the existence of gases.

27 Alexander Ross, quoted in G. Keynes edition of Sir Thomas

Browne's collected works (first edition 1686, Keynes edition 1964).

28 Jan Swammerdam (1637–1680), Dutch naturalist, who wrote a classic of microscopy.

28 Antonio Vallisnieri (1661–1730), Italian physician and naturalist, studied anatomy and ontogeny of animals.

28 Anton van Leeuwenhoek (1632–1723), Dutch naturalist.

28 John Turberville Needham (1713–1781), English naturalist, advocate of spontaneous generation.

28 William Harvey (1578–1657), English physician who is considered to be one of the founders of modern medicine.

28 Georges-Louis Leclerc, compte de Buffon (1707–1788), French naturalist.

28 Lazzaro Spallanzani (1729–1799), Italian physiologist, opponent of the theory of "vital atoms." He studied regeneration and transplantation in lower animals.

28 Max Schultze (1825–1874), German cytologist and histologist who discovered the protoplasm and nucleus of the cell.

28 Theodor Schwann (1810–1882), German histologist and physiologist who advanced the cell theory for plants and isolated pepsin.

29 Félix-Archimède Pouchet (1800–1872), French naturalist who thought that organisms can be produced by fermentation or putrefaction.

29 Louis Pasteur (1822–1895), French chemist and microbiologist who discovered the processes of fermentation and vaccination.

29 Paul Busse-Grawitz (1850–1933), German pathologist who fabricated the theory of "dormant cells" without nuclei.

Off-the-Wall Inventions

31 Thomas H. Holmes (1817–1900), U.S. physician, inventor of embalming devices.

31 John Dilks, quoted in Christopher Clemens and Mark Smith, *Death, Grim Realities and Comic Relief* (New York: Delacorte Press, 1982).

33 Alfred Jarry (1873–1907), French satirist playwright who invented "pataphysics," a logic of the absurd.

33 Marcel Duchamp (1887–1968), French (and later American) artist, one of the pioneers of Dada.

33 Charles Babbage (1792–1871), English mathematician.

34 Johann von Neumann (1903–1957), German-American mathematician who made important contributions to quantum physics and computer science.

34 John Presper Eckert Jr. (1919–1995), American engineer; coinventor, with John W. Mauchly, professor of electrical engineering, of the first general purpose electronic computer.

What Links Me with Ladislaus the Posthumous

37 Quotations are from a fifteenth-century Czech chronicle; see Gustav Gellner, *The Disease of Ladislav Pohrobek* (Prague: CCH, 1934).

38 Frantisek Palacký (1798–1876), historian, one of the leading figures of Czech cultural rebirth in the nineteenth century.

39 Kanter and Lewin, quoted in Gustav Gellner, *The Disease of Ladislav Pohrobek* (Prague: CCH, 1934).

42 Dr. Emanuel Vlcek, Czech anthropologist specializing in pathographies of Czech rulers.

Can Man Create His Double?

45 Karel Capek, *The War with the Newts* (Czech edition, 1936); New York: Bantam, 1964).

48 Lewis Thomas, "On Cloning a Human Being," in *The Medusa and the Snail: More Notes of a Biology Watcher* (New York: Viking Press, 1979).

49 Geryon is a three-bodied giant in Greek mythology who was killed by Hercules.

Apes, in Particular

50 Durs Grünbein, "Ein Schimpansen im Londoner Zoo" (A Chimpanzee in the London Zoo), in *Falten und Fallen* (Frankfurt am Main: Suhrkamp, 1994).

52 A golem, in Jewish folklore, is a clay figure constructed in the form of a human being and endowed with life.

On Pigs As a Laboratory Vacuum

55 Antonin Artaud (1896–1948), French playwright who envisioned a theater of cruelty.

59 Alfred Brehm, *Brehms Tierleben,* vol. 1, *Die Saugtiere* (Leipzig and Vienna: Bibliographisches Institut, 1911–19).

The Electron Microscope Tesla 242 D

61 Miles, Gurr, Zeiss Wetzlar, LKB, suppliers of chemicals and scientific instruments.

The Paranoid Nymphs

66 Antoine de Saint-Exupéry (1900–1944), French pilot and author, in *Flight to Arras* (London: Heinemann, 1942); originally published in French as *Pilote de guerre.*

66 R. O. Finlay and P. T. Rudd, "Current Concepts of the Aetiology of SIDS," *British Journal of Hospital Medicine* 49: 727–32 (1993).

From the Amoeba to the Philosopher

69 Ernest Rutherford (1871–1937), British physicist who worked on radioactivity and produced a model of the atom.

70 Paul Feyerabend, "How to Defend Society against Science," in *Scientific Revolutions,* ed. I. Hacking (New York: Oxford University Press, 1981).

70 Tadeusz Kotarbinski (1886–1970), Polish philospher, cofounder of semantics.

70 G. M. Boshyan, O. B. Lepeshinskaya, Russian biologists after World War II who were blindly loyal to the ruling communist party.

70 Trofim Denisovich Lysenko (1898–1976), Russian biologist and agronomist who spearheaded the Stalinist opposition to modern genetics.

73 Peter B. Medawar, *Pluto's Republic* (Oxford and New York: Oxford University Press, 1982).
73 Thomas Hobbes (1588–1679), English philosopher; *Leviathan* was published in 1651.
73 Francis Bacon (1561–1626), English philosopher and statesman, in *Novum Organum* (1620).
73 William Godwin (1756–1836), English political writer, in *An Enquiry Concerning Political Justice* (London, 1793).
73 Stanislaw Jerzy Lec, *Unkempt Thoughts* (1959). Translated by Jacek Galazka (New York: St. Martin's Press, 1962).
74 James Thurber (1894–1961), American humorist writer.
74 George A. L. Sarton, *The Study of History of Science* (New York: Dover, 1957).
74 Karl R. Popper, "The Rationality of Scientific Revolutions," in *Scientific Revolutions,* ed. I. Hacking (New York: Oxford University Press, 1981).
75 Ralph Estling, "Universal Darwinism," *Nature* (London) 361: 489 (1993).
75 Jacques Monod (1910–1976), French biochemist, in *Le hasard et la nécessité* (Chance and Necessity) (Paris: Seuil, 1970).
75 Stephen Jay Gould, *The Flamingo's Smile* (New York: Norton, 1985).
75 —. *Full House* (New York: Norton, 1996).
76 —. *Wonderful Life* (New York: Norton, 1989).

Trouble on the Spaceship

81 R. Buckminster Fuller, *Operating Manual for Spaceship Earth* (New York: Simon and Schuster, 1969).
82 Stephen Jay Gould, *Eight Little Piggies* (New York: Norton, 1993).
85 Peter Medawar, in *Pluto's Republic* (Oxford and New York: Oxford University Press, 1982).
85 Georg Christoph Lichtenberg (1742–1799), German physicist and satirist.

A Journey to Jupiter

86 Fred Hoyle's Black Cloud is a science fiction image of a conscious interstellar gas cloud, traveling among the stars and feeding off free energy.

87 Stephen Hawking (1942–), British physicist and cosmologist, author of *A Brief History of Time.*

88 Alan Lightman, *A Modern Day Yankee in a Connecticut Court* (New York: Viking, 1986).

91 The Arecibo Observatory in Puerto Rico is the world's largest single-unit radio telescope.

The Night Song

94 Proclus Diadochus (410–485), Greek Neoplatonist philosopher who wrote on astronomy and mathematics.

What the Nose Knows

99 Karel Siktanc (1928–), Czech poet, in *Czech Horologe* (Cesky orloj) (Praha: Prace, 1990).

103 Premysl Orac (the Plowman), mythical founder of the Premyslid dynasty, which ruled Czech lands from about 800 to 1306.

By Nature Alone

104 Jean-Jacques Rousseau (1712–1778), French philosopher and writer, in *Émile; ou, de l'éducation* (1762). Passage translated by Miroslav Holub.

105 Jean-Baptiste Molière (1622–1673), French playwright.

105 Anaximenes (died ca. 500 B.C.), Ionic philosopher.

105 Sir Isaac Newton (1642–1727), English scientist and mathematician.

106 Friedrich Wöhler (1800–1882), German chemist, discoverer of aluminum.

107 Roald Hoffmann (1937–), Polish-American chemist and writer of popular science books, including *The Same and Not the Same* (New York: Columbia University Press, 1995).

111 William Carlos Williams (1883–1963), American physician and poet.

Symbiotic Tranquility

112 Gaia, Greek goddess of earth. The British chemist James Lovelock uses Gaia as a metaphor for the coevolution of climate and life.

114 Edgar Allan Poe (1809–1849), American poet and short story writer.

114 Robert Lowell, "Waking Early Sunday Morning," in *Near the Ocean* (New York: Farrar, Straus and Giroux, 1967).

Beasts and Freaks

118 James Thurber, "Less Alarming Creatures," in *The Beast in Me and Other Animals* (New York: Avon Books, 1948).

118 Lewis Thomas, "Some Biomythology," in *The Lives of a Cell* (New York: Viking, 1975).

120 Bozena Nemcová (1820–1862), founder of the new Czech prose who rewrote many Czech fairy tales.

120 J. R. R. Tolkien (1892–1973), English writer, author of *The Hobbit* and *The Lord of the Rings*.

The Death of Butterflies

122 Frána Srámek (1877–1952), Czech poet and playwright.

123 Jaroslav Seifert (1901–1986), Czech poet who won the Nobel Prize in 1984.

123 Sir Vincent B. Wigglesworth (1899–1994), British entomologist and founder of modern insect endocrinology.

123 William Blake (1757–1827), English poet, "Auguries of Innocence" in *Songs of Experience* (1794).

123 Lincoln Brower, American entomologist and lepidopterist.

126 William Blake, "The Sick Rose" in *Songs of Experience* (1794).

Otters, Beavers, and Me

128 Lewis Thomas, *The Medusa and the Snail: More Notes of a Biology Watcher* (New York: Viking, 1979).

130 Bedrich Smetana (1824–1884), Czech composer and founder of the Czech National Opera.

131 Jean-Paul Sartre (1905–1980), French existentialist philosopher, dramatist, and novelist.

Shedding Life

136 Keres, goddesses of violent death in Greek mythology.

Unter den Linden

140 Jean Arp (1887–1966), French artist and poet.

141 Jaroslav Hasek, *The Good Soldier Schweik* (1920–23), translated by Cecil Parrott (New York: Viking Penguin, 1985).

141 Eugène Ionesco (1912–), French-Romanian playwright of the absurd.

141 Galileo Galilei (1564–1642), Italian astronomer and physicist.

141 Thales of Miletus (625?–547 B.C.), Greek philosopher.

141 Immanuel Kant (1724–1804), German philosopher.

142 R. S. Root-Bernstein, *Discovering: Inventing and Solving Problems at the Frontiers of Scientific Knowledge* (Cambridge, Mass.: Harvard University Press, 1989).

142 Nikolay Ivanovich Lobachevsky (1792–1856), Russian mathematician, cofounder of non-Euclidean geometry.

The Discovery: An Autopsy

144 Diego de Estella (1524–1578), quoted by Robert King Merton in *On the Shoulders of Giants* (New York: Free Press, 1965).

144 James D. Watson, *The Double Helix: A Personal Account of the Discovery of the Structures of DNA* (New York: Atheneum, 1968).

145 Robert K. Merton, *The Sociology of Science* (Chicago: University of Chicago Press, 1973).

145 Frederic L. Holmes, *Claude Bernard and Animal Chemistry.* (Cambridge: Harvard University Press, 1974).

145 —. *Hans Krebs: The Formation of Scientific Life* (New York: Oxford University Press, 1991).

145 D. Kulkarni and H. A. Simon, "The Process of Scientific

Discovery: The Strategy of Experimentation." *Cognitive Science* 12, no. 2 (1988): 139–75.

145 Henri Bergson (1859–1941), French philosopher who won the Nobel Prize for Literature in 1928. He is the author of *Laughter: An Essay on the Significance of the Comic* (1900) and *Creative Evolution* (1907).

146 Milton Rothman, *The Science Gap: Dispelling the Myths and Understanding the Reality of Science* (Buffalo: Prometheus Books, 1992).

146 J. Sasso, "The Stages of the Creative Process," *Proceedings of the American Philosophical Society* 124 (1980): 119–32.

146 Joseph S. Fruton, *A Skeptical Biochemist* (Cambridge, Mass.: Harvard University Press, 1992).

147 Thomas S. Kuhn, "Logic of Discovery or Psychology of Research" in *The Essential Tension* (Chicago: University of Chicago Press, 1977).

Science and the Corrosion of the Soul

154 Bryan Appleyard, *Understanding the Present: Science and the Soul of Modern Man* (London: Picador, 1992).

154 Francis Fukuyama, *The End of History and the Last Man* (New York: Free Press, 1992).

155 Edmund Husserl (1859–1938), German philosopher, founder of phenomenology, who in *Crisis of European Sciences* (1936) formulated an argument on the gap between our "natural" world and the exact and formalized world of sciences.

155 Ladislav Kovac, "Education in the Light of Cognitive Biology (in Slovakian)," *Vesmir* 74 (1995): 252–54.

157 Richard Rorty, *Philosophy and the Mirror of Nature* (Princeton: Princeton University Press, 1979).

157 Mario Vargas Llosa (1936–), Peruvian novelist, essayist, and playwright.

158 Milton Rothman, *The Science Gap: Dispelling the Myths and Understanding the Reality of Science* (Buffalo, N.Y.: Prometheus, 1992).

159 Dannie Abse, "Seekers after the Truth," in *Collected Poems* (London: Hutchinson, 1977).

159 Vilém Laufberger (1890–1977), Czech physiologist, known for the successful induction of metamorphosis in larval salamanders.

159 Harold Bloom, *Agon: Towards a Theory of Revisionism* (New York: Oxford University Press, 1982).

160 Lewis Thomas, "Autonomy," in *The Lives of a Cell: Notes of a Biology Watcher* (New York: Viking, 1974).

161 Schrödinger's cat is a thought experiment illustrating the basis of quantum mechanics. Until it is seen by an observer, the cat exists in some indeterminate state.

163 Jacob Bronowski (1908–1974), Polish-British mathematician and philosopher of science.

164 Norbert Bischof, *Gescheiter als alle die Laffen* (Munich: Piper & Co., 1993).

165 W. H. Auden (1907–1973), English poet.

Whatever the Circumstances

173 Pierre Janet (1859–1947), French psychologist and precursor of Freud.

Science and Morality

175 Jacob Bronowski, *Science and Human Values* (New York: Harper and Row, 1965).

176 Richard P. Feynman, *The Feynman Lectures on Physics,* 1 (London: Addison-Wesley, 1963).

176 Thomas S. Kuhn (1922–), professor of philosophy and history of science, author of *The Structure of Scientific Revolutions* (Chicago: University of Chicago Press, 1970).

176 Pierre-Simon Laplace (1749–1827), French mathematician and astronomer who applied Newtonian theory of gravity to the solar system and published influential books on celestial mechanics.

177 Maxwell's demon, imagined by J. C. Maxwell in 1871 as a hypothetical being capable of detecting and separating slow and fast molecules, thus "creating" energy.

178 Johann Wolfgang von Goethe (1749–1832), German poet, dramatist, and scientist.

179 "Oblomov attitude" refers to Count Oblomov, the protagonist of a novel by I. A. Goncharov (1812–1891) about a passive and lazy nobleman who never achieves anything.

181 Søren Kierkegaard (1813–1855), Danish religious philosopher, critic of rationalism, regarded as a founder of existentialism.

181 Victor F. Weisskopf, "Why Pure Science?" in *Bulletin of the Atomic Scientists* 21 (1965): 4–8.

183 Aleksandr I. Herzen (1812–1870), Russian writer and liberal philosopher, exiled in 1847.

Wisdom As a Metaphor

184 Karel Capek, "Agathon; or, On Wisdom" (1920). Passage translated by Miroslav Holub.

188 Bertolt Brecht (1898–1956), *Galileo,* trans. by Charles Laughton (New York: Grove Weidenfeld, 1966).

188 Stansilav Komárek, in *Nová Prítomnost* (1996). Passage translated by Miroslav Holub

189 V. G. Belinsky (1811–1848), influential Russian literary critic.

191 Charles Péguy (1873–1914), French poet and essayist.

192 William Shakespeare, *All's Well That Ends Well* 2.3.

195 Lynn Margulis, "Symbiosis and Evolution," *Scientific American* 225 (1971).

196 Alfred North Whitehead (1861–1947), English mathematician and philosopher.

196 François Jacob (1920–), French biologist who described the translation mechanism of protein synthesis.

197 Julian Tuwim (1894–1953), Polish poet and satirist.

199 Richard Kearney, Irish philosopher, author of *Poetics of Imagining from Husserl to Lyotard* (1991). Quotation retranslated from the Czech by Miroslav Holub.

200 Albert Camus (1913–1960), French existentialist novelist and essayist.

A Concert in Morelia

204 Carl Sagan, *Broca's Brain: Reflections on the Romance of Science* (New York: Random House, 1979).

A Scientific Horror Story

209 W. J. Broad and N. Wade, *Betrayers of the Truth* (New York: Simon and Schuster, 1982).

The Kenrak System

212 Johann Gottlieb Fichte (1762–1814), German philosopher, follower of Kant, and one of the leading forces of European moral philosophy.

Sukhumi; or, Recollections of the Future

219 Sukhumi is the capital of the Abkhaz Republic on the Black Sea.

In Search of the Enemy

233 Jan Patocka (1907–1977), Czech phenomenologist, pupil of Husserl, hero of the anticommunist resistance after the Russian invasion in 1968.

233 Jan Palach, college student who burned himself to death in Wenceslas Square in 1969 in protest of the Soviet occupation of Czechoslovakia.

In Republic Square

242 Konrad Lorenz, *On Aggression* (New York: Harcourt, Brace and World, 1966).

243 Erich Fromm (1900–1980), German-born American philosopher and psychoanalyst, author of *The Sane Society* (1955) and *The Revolution of Hope* (1968).

Slavery and Worse

245 Diocletian (245–313), Roman emperor.

245 Harold Morowitz, *The Wine of Life and Other Essays on Societies, Energy and Living Things* (New York: St. Martin's Press, 1979).

246 Habeas Corpus Act (1679), English law defining the conditions of arrest so as to prevent arbitrary imprisonment.

248 Aaron Sachs, "The Last Commodity: Child Prostitution in the Developing World," *World Watch,* July 4, 1994.

No

256 Karl Kraus (1874–1936), Czech-Austrian satirist, journalist, and poet.

256 Rabindranath Tagore (1861–1941), Indian poet.

260 Ivan P. Pavlov (1849–1936), Russian physiologist who wrote a classic work on conditioned reflexes.

262 Jan Werich (1905–1980), Czech actor and writer.

CONVERSION TABLE

1 kilometer = 0.62 miles

1 meter = 39.4 inches

1 liter = 1.6 quarts

0°K (absolute zero) = -273.15°C = -459.67°F

A Note on the Translations

All of these essays were edited by David Young, in consultation with Miroslav Holub. Their first English versions, however, were produced by five different individuals: Dana Hábova, Vera Orac, Todd Morath, Catarina Vocadlova, and the author, as follows:

Hábova/Patricia Debney/David Young: "Shedding Life"

Dana Hábova: "Apes, in Particular," "Otters, Beavers, and Me," "Unter den Linden," "Science and Morality"

Vera Orac: "This Long Disease," "Kidneys and History," "On Pigs As a Laboratory Vacuum," "The Electron Microscope Tesla 242 D," "What the Nose Knows," "By Nature Alone," "The Kenrak System," "Red Noodles; or, About Uselessness"

Todd Morath: "The Experiment of 1668," "Off-the-Wall Inventions," "A Scientific Horror Story"

Catarina Vocadlova: "The Death of Butterflies," "The Discovery: An Autopsy," "Sukhumi; or, Recollections of the Future"

Miroslav Holub: "Zen and the Thymus," "What Links Me with Ladislaus the Posthumous," "Can Man Create His Double?" "The Paranoid Nymphs," "From the Amoeba to the Philosopher," "Trouble on the Spaceship," "A Journey to Jupiter," "The Night Song," "Symbiotic Tranquility," "Science and the Corrosion of the Soul," "Whatever the Circumstances," "Eureka," "In Search of the Enemy," "In Republic Square," "Slavery and Worse," "No"

Although MIROSLAV HOLUB's writings were banned for twelve years after the Russian occupation in his homeland of Czechoslovakia, this doctor of medicine and philosophy has been translated and read worldwide.

Born in 1923 in Pilsen, Western Bohemia, Holub studied science and medicine at Charles University, Prague, where he worked in the departments of philosophy and the history of science, as well as in a psychiatric ward. Holub supported himself as an editor of the scientific magazine, *Vesmìr,* at a time when he had also begun to write poetry. He earned an M.D. degree in 1953 and a specialization in pathology, which eventually led him to join the immunological section of the Biological, later Microbiological Institute of the Czechoslovak Academy of Science in 1958. That same year he earned his Ph.D. and published his first book of poems.

As a poet, Holub became associated with the literary group Poetry of Everyday, which existed until the Russian occupation in 1968 to 1970 (when most writers became nonpersons). From 1965 to 1967 he also worked as an immunologist in New York and in Freiburg, Germany, in 1968.

Holub is the author of over 170 scientific papers, sixteen books of poems, and ten books of essays.

Interior design by Wesley B. Tanner
Typeset in Sabon by Stanton Publication Services, Inc.
Printed on acid-free Liberty paper by Quebecor Fairfield

More nonfiction from Milkweed Editions

Changing the Bully Who Rules the World: Reading and Thinking about Ethics
Carol Bly

The Passionate, Accurate Story: Making Your Heart's Truth into Literature
Carol Bly

Transforming a Rape Culture
Edited by Emilie Buchwald, Pamela Fletcher, and Martha Roth

Rooms in the House of Stone
Michael Dorris

The Most Wonderful Books: Writers on Discovering the Pleasures of Reading
Edited by Michael Dorris and Emilie Buchwald

Boundary Waters: The Grace of the Wild
Paul Gruchow

Grass Roots: The Universe of Home
Paul Gruchow

The Mythic Family
Judith Guest

The Art of Writing: Lu Chi's Wen Fu
Translated from the Chinese by Sam Hamill

Chasing Hellhounds: A Teacher Learns from His Students
Marvin Hoffman

Coming Home Crazy: An Alphabet of China Essays
Bill Holm

The Heart Can Be Filled Anywhere on Earth:
Minneota, Minnesota
Bill Holm

Rescuing Little Roundhead
Syl Jones

I Won't Learn from You! The Role of Assent in Learning
Herbert Kohl

Basic Needs: A Year with Street Kids in a City School
Julie Landsman

Tips for Creating a Manageable Classroom:
Understanding Your Students' Basic Needs
Julie Landsman

Planning to Stay:
Learning to See the Physical Features of Your Neighborhood
William R. Morrish and Catherine R. Brown

The Old Bridge:
The Third Balkan War and the Age of the Refugee
Christopher Merrill

A Male Grief: Notes on Pornography and Addiction
David Mura

Homestead
Annick Smith

What Makes Pornography "Sexy"?
John Stoltenberg

Testimony:
Writers of the West Speak On Behalf of Utah Wilderness
Compiled by Steve Trimble and Terry Tempest Williams